T0106591

Goal
Question
Metric
Stakeholders
A MEASUREMENT
FRAMEWORK FOR
SOFTWARE PROJECTS
A Generic and Practical Goal-Question-
Metric (GQM) based approach
Prashanth Harish Southekal

A MEASUREMENT FRAMEWORK FOR SOFTWARE PROJECTS

A Generic and Practical Goal-Question-Metric (GQM) based approach

PRASHANTH HARISH SOUTHEKAL

Order this book online at www.trafford.com
or email orders@trafford.com

Most Trafford titles are also available at major online book retailers.

Printed in the United States of America.

ISBN: 978-1-4269-8180-7 (sc)
ISBN: 978-1-4269-8181-4 (e)

Trafford rev. 08/02/2011

www.trafford.com

North America & international
toll-free: 1 888 232 4444 (USA & Canada)
phone: 250 383 6864 • fax: 812 355 4082

To my family.

Acknowledgements

This book is my PhD thesis which I undertook to address the acute problem of software project visibility for the project stakeholders which has ultimately resulted in the poor failure rate of software projects.

This book attempts to bring together the software metrics researchers and industry practitioners with a generic and practical Goal-Question-Metric (GQM) based measurement framework for software projects.

In this PhD journey for over three years, there have been many people who have positively impacted this "PhD project". I would like to thank them for their role in the completion of the doctoral thesis on "Formulation and Validation of a Generic Goal-Question-Metric (GQM) Based Measurement Framework for Software Projects". Pursuing the PhD program was a unique learning and collaborative experience and it has definitely been one of my best "investments" till date.

First and foremost, I offer my sincerest gratitude to my supervisors - Dr Ginger Levin and Dr Darren Dalcher, who have supported me throughout my thesis with their knowledge, rigor and patience whilst providing me the space to experiment and work according to my style and convenience. While Dr. Levin looked at the finer details in my research and always kept me motivated to focus on my work, Dr. Dalcher looked at the fundamental aspects in the research ensuring that I have a proper "story" to tell. It has been my good fortune to gain such an incredible knowledge and exposure by working with these two movers and shapers of project management. I simply could not wish for better or friendlier supervisors as they have set an excellent example not just as successful professors and mentors but also as wonderful human beings.

I also had the privilege to be associated with some renowned figures of project management in Skema. Professor Christophe N Bredillet introduced me to the PhD program and provided all the guidance needed to kick-start the doctoral work. Professor Rodney Turner understood my research topic and identified the right supervisors to guide me. Professors Ralf Muller and Philippe Ruiz shared their valuable insights on the business research methods. It has been a great honor for me to be their student and pursue research reflecting the current trends of globalization in software project management – an Indian student living in Canada, working for a global company, pursuing PhD in a French university, supervised by professors from the United States and United Kingdom.

In the entire research period, I had numerous discussions with top researchers and industry practitioners who were instrumental in giving a better shape to this research. These researchers include Capers Jones (Software Productivity Council, US), Professors Guenther Ruhe and Yingxu Wang (University of Calgary, Canada), Professors Giuliano Antoniol (University of Montreal, Canada) , Professor Ken Caner (Florida Institute of Technology, US) , Dr KC Shashidhar (Max Planck Institute, Germany), Tom Gilb (Result Planning Limited) and Professor PP Iyer (Indian Institute of Science, India). Industry practitioners who helped me in this endeavor include Dr. Stephane Vaucher (Metrics Specialist, Benchmark Consulting, Canada), Michael Roomey (Master Black Belt, General Electric, US), Liam Durbin (Ex CIO, General Electric, US), Ray Stratton (President, Management Technologies, US), Abhay Shetty (Project Manager, SAP, India), members of the LinkedIn group “Measurement and Analysis Forum” and all the respondents who participated in the survey and poll.

I am extremely grateful to my current employer Accenture Inc and my former employer SAP AG for providing me a conducive and stimulating work atmosphere for research. I am also grateful to ATB Financial, Canada, for giving me an opportunity to implement this research in the Core SAP banking program. Implementing the measurement framework was a team work and I would sincerely thank my Core SAP Banking Channel stream colleagues –Vijay Kannan (SAP-India) and Kishor Gopinathan (SAP-India) for their help in data collection and inclusiveness in the implementation of the measurement framework. I also acknowledge the efforts of Desmond Tsui (Accenture Canada) for taking the pains in editing the content and suggesting improvements to the thesis. My sincere thanks also go to the Skema support staff that made every effort to make my stay on campus a comfortable one.

I am also thankful to my parents, my in-laws, and my sister and her family for all their help, blessings and wishes. Finally, I owe my loving thanks to my wife Shruthi who has supported and encouraged me in all my pursuits and my two wonderful kids Pranathi "Panna" and Prathik "Heera" for making my life a celebration.

Prashanth Harish Southekal
Calgary, AB, Canada
August 2011.

Table of Contents

List of Tables

List of Figures

List of Key Acronyms and Abbreviations

- AC – Actual Cost
- BSC – Balanced Scorecard
- CA – Control Account
- CAPEX – Capital Expenditures
- COQ – Cost of Quality
- COTS – Commercial-Off-The-Shelf
- CSF – Critical Success Factors
- CMMI - Capability Maturity Model Integration
- Cpk – Sigma Level
- CPI – Cost Performance Index
- CV – Cost Variance
- DD – Defect Density
- DPMO - Defects Per Million Opportunities
- DRE – Defect Removal Efficiency
- EAC – Estimate At Completion
- ERP – Enterprise Resource Planning
- ES – Earned Schedule
- ESE – Empirical Software Engineering
- EV – Earned Value
- EVM – Earned Value Management
- FP – Function Point
- GQM – Goal Question Metric
- HSS - Halstead's software science
- ISO - International Organization for Standardization
- IT - Information Technology
- JAD – Joint Application Development
- KPI - Key Performance Indicator
- KW – Kruskal Wallis Test
- (K)LOC – (Kilo) Lines of Code

- OBS – Organizational Breakdown Structure
- OLTP – Online Transaction Processing System
- OLAP – Online Analytical Processing System
- PCA – Paired Comparison Analysis
- PMB – Performance Measurement Baseline
- PMBOK – Project Management Body of Knowledge
- PMI - Project Management Institute
- PSM – Practical Software Measurement
- PV - Planned Value
- RBS – Requirement Breakdown Structure
- ROI - Return on Investment
- SAAS – Software As A Service
- SDLC – Software Development Lifecycle
- SEI – Software Engineering Institute
- SPI – Schedule Performance Index
- SPI - Software Process Improvement
- SRS - Software Requirement Specification
- SS – Sample size
- SV – Schedule Variance
- TCO - Total Cost of Ownership
- TO – Total Opportunities
- V(G) – McCabe's Cyclomatic Complexity
- VOC – Voice of the Customer
- VOP – Voice of the Process
- WBS – Work Breakdown Structure
- WP – Work Package

Abstract

The Standish Group's "CHAOS Summary 2009" report asserts that just one in three software projects are delivered on time, within budget, with the required features and functions. While it would appear that this "Software Crisis" is primarily caused by the inherent complexity in the projects, one of the approaches to ameliorate the poor success rate in software projects is formulating a stakeholder driven measurement framework (in a project status report) for taking suitable corrective actions at the right time. In this backdrop, the challenge is to derive a generic and objective measurement framework that is validated theoretically and empirically reflecting the goals of the project stakeholders – initiators, implementers and beneficiaries.

Based on the Goal-Question-Metric (GQM) model, this research derived a generic objective measurement framework using six generic steps that addresses the concerns of the stakeholders in software projects. The measurement framework includes eight measures formulated from 12 questions related to five attributes namely size, complexity, schedule, cost and quality. The eight measures are Lines of Code (LOC), Function Points (FPs), McCabe's Cyclomatic Complexity (VG), Schedule Performance Index (SPI), Cost Performance Index (CPI), Sigma level (Cpk), Defect Density (DD) and Defect Removal Efficiency (DRE). The project status report with these eight measures is meant to be a succinct summary capturing critical information most relevant to the three groups of stake holders. However depending on the unique project circumstances more measures can be suitably added on top of this measurement framework.

The validation of this measurement framework rests on the theoretical and empirical validation of two hypotheses.

1. The eight measures are the best measures to derive the objective status of their respective attributes.

2. The eight measures can serve as a generic core set for an accurate and objective status of a software project.

To validate these two hypotheses theoretically, seven criteria from measurement theory were applied to ensure that the each measure characterizes the property it claims to measure. The objective of theoretical validation is to minimize the chances of failure encountered during the implementation of the measurement framework by addressing exceptions. At the measurement framework level, the ten questions of Cem Kaner and Walter Bond [Kaner and Bond, 2004] were also applied for compliance.

Though theoretical validation using measurement theory is getting a great deal of attention from researchers, industry practitioners still rely on empirical evidence of a measure's utility. Hence this research empirically validated the two hypotheses using eight criteria with a statistical analysis of data obtained via survey and case studies (controlled and uncontrolled). A survey instrument using the Likert response scale was implemented. Based on 110 stakeholders from 29 countries sharing more than 1750 years of their software project experience, these eight measures were found to be a positive predictor of project success. The measurement framework was further tested by implementing it in a real world software project (controlled setting) to measure the project for its successful completion. Also another project (uncontrolled setting) which did not use this measurement framework failed as the stakeholders did not have the accurate and complete information of the project. The survey and the two case studies strengthen the hypothesis empirically that the proposed measurement framework when properly implemented provided critical information at the right time for proactive decision making.

Chapter 1: Introduction - The Problem Statement and Research Proposition

1.1 The Changing Landscape of Software Project Management

Software project management encompasses the knowledge, techniques, and tools necessary to manage the development of software products involving multiple disciplines [Tomayko and Hallman, 1989]. It is also the art of balancing competing objectives, managing risk, and overcoming constraints to deliver successfully the software product satisfying the needs of the customers. As good project implementation begins with the understanding of the project attributes, a software project has seven main attributes [**Gido and Clements, 1998**]:

1. **Purpose.** There must be a purpose to justify the need and existence of a project.

2. **Length.** Project Management Body of Knowledge (PMBOK) describes a project as "a unique temporary endeavor" [PMBOK, 2008]. In other words, a software project has set beginning and end dates mainly derived goals, scope and complexity.

3. **Ambiguity.** Every software project will face some degree of ambiguity primarily driven from complexity and uncertainty. Factors such as schedule, resources and budgets can change due to unforeseen circumstances beyond the control of the project.

4. **Customers.** They are interested parties who have a stake in a project and for whom the project is implemented.

5. **Resources.** In a software project resources can include skilled employees, hardware, software, budgets and any other assets as deemed necessary.

6. **Unique Endeavour**. Every software project is unique because they have never before been attempted before.

7. **Interdependent tasks**. A project is always related to various tasks and these tasks need to be accomplished in a certain sequence in order to accomplish project's objective.

But in the recent years the landscape of software project management has undergone a significant metamorphosis around the globe. Software solutions are changing to be user centric, web centric, service oriented and implemented through new delivery models. In this scenario, five major trends are reshaping the field of software project management [Gartner, 2008].

Trend 1: Use of mass collaboration tools

The demand for "cheaper-faster-better" software, combined with Wikinomics – the use of mass collaboration in a business environment driven by companies such as Google, Wikipedia, Facebook, LinkedIn, Twitter, YouTube and other collaborative and networking tools have dramatically changed the way people work and behave. Although technology companies are leading the way, their efforts are helping other traditional companies such as Unilever and GE, in directly collaborating with end customers and partners in creating products and services faster, with fewer defects, at far lower costs and risk. In addition, "information glut" a common term only a few years back, is no longer discussed indicating a significant change in the information handling management capabilities of the IT users. The Gartner group believes that by 2015 no company will be able to build or sustain a competitive advantage unless it capitalizes on the combined power of individualized behavior, social dynamics and collaboration [Gartner, 2008]. These changes have basically enhanced the expectations of the users from software projects.

Trend 2: Changes in the IT Eco-System

In the last five years the enterprise application market has become consolidated – both from the products and services front. On the products side, Intel has acquired McAfee. HP purchased Palm. Oracle acquired

ATG, Seibel and Sun Microsystems. SAP has acquired Business Objects and Sybase. These events have resulted in the convergence of most enterprise applications with a few big companies who provide a host of refined and tested functionalities for most users through their Commercial-off-the-shelf (COTS) products. On the IT services front, recent acquisition examples include Affiliated Computer Services (ACS) by Xerox, EDS by HP and Perot Systems by Dell. This "enterprise consolidation" has resulted in the emergence of mega-vendors who are seeking to dominate the enterprise architecture with their ecosystems which include not only the hardware and software, but also developers, applications, advertising, location-based services, unified communications to name a few. The end result is reduced "bargaining power" for organizations which implement software projects with regards to the technology options.

Trend 3: Increasing Complexity

Powerful business forces such as regulatory compliance, information management, government oversight, changing operating models due to mergers and acquisitions, speed of innovation, competition, and tax policies have reshaped the global business landscape. These developments have resulted in complexity as a source of challenge, change, risk, unpredictability and even opportunity in some instances. On the technical front, the IT infrastructure today is typically made up of a mosaic of different systems from different vendors integrated through ESB (Enterprise Service Bus) and SOA (Service Oriented Architecture) integration. The mosaic of technologies, frameworks and platforms make e-business implementation, Cloud computing enabled solutions, Data Center Management, Legacy Integration etc are not only rich in semantics from a domain perspective but are also complex to assimilate. This complexity is creating a critical need for new skills (technical and managerial) for successful project delivery.

Trend 4: Spirit of SaaS

Software as a Service (SaaS), commonly referred to as the Application Service Provider (ASP) model, is heralded by many as the game-changer in application software arena. Following the maxim that "the Internet changes everything," many believe that Commercial-off-the-shelf (COTS) applications will soon be swept away by Web-based, outsourced products and services. In fact, SIIA (Software and Information Industry Association) believes that the SaaS model is capable of causing a sea change in the software industry [SIIA, 2001]. While such drastic predictions have not yet come true, because of to technical and business reasons, the spirit of this thought and change is affecting all stakeholders not just in software projects but also in the entire software industry.

Trend 5: Power Shift

There is an increasing emphasis on lines of business (LOB) to have direct access to customers and IT functions. Power has shifted to the business owners who now make decisions on software projects. In addition, given that knowledge is playing a key role in today's information age business, outsourcing/distributed delivery is transforming the entire industry and changing the way business is done. Companies are looking at speed by leveraging the time zone differences and utilizing low cost talent available globally to deliver projects. The most important asset in an information age company is turning out to be knowledge, not capital. The underlying message from this trend is that power in the company has now shifted to knowledge workers and they hold key for successful software projects.

These five major trends have radically increased the complexity in software project management given that complexity is considered a key factor in the poor success rate of software projects [Tatikonda and Rosenthal, 2000]. Software project management is not just a technical issue which can be resolved with technology; it is a multi-disciplinary business engagement, requiring ownership, commitment, understanding and alignment of business and IT strategies within an organization.

1.2 Software Projects and Evolution

Software project management enables translation of a problem statement into a solution that satisfies the needs of all the stakeholders. Software projects have evolved over years demonstrating different characteristics and demanding different management approaches. Lehman et al who have studied this evolution from 1974 have come up with eight laws of software evolution for real world systems describing a balance between forces driving new developments on one hand, and forces that slow down progress on the other hand [Lehman, Ramil, Wernick, Perry and Turski, 1997]. The eight laws of software evolution are:

1. **Continuing Change (1974).** Systems must be continually adapted or they become progressively less satisfactory.

2. **Increasing Complexity (1974).** As systems evolve their complexity increases unless dedicated work is done to maintain or reduce it.

3. **Self Regulation (1974).** System evolution process is self regulating with distribution of product and process measures close to normal.

4. **Conservation of Organizational Stability (1978).** The average effective global activity rate in an evolving system is invariant over product lifetime.

5. **Conservation of Familiarity (1978).** As the system evolves all stakeholders associated with it such as developers, quality assurance staff, users must maintain mastery of its content and behavior to achieve satisfactory evolution.

6. **Continuing Growth (1991).** The functional content of the systems must be continually increased to maintain user satisfaction.

7. **Declining Quality (1996).** The quality of the system will decline unless it is rigorously maintained and adapted to operational environment changes.

8. **Feedback System (1996).** The system evolution processes constitute multi-level, multi-loop, multi-agent feedback systems and must be treated as such to achieve significant improvement.

In this setting, the software projects that typically run inside an organization can be classified into three types whether it is Online Transactional Processing Systems (OLPTs), Online Analytical Processing Systems (OLAP), Web Applications etc.

1. Commercial-off-the-Shelf (COTS) Projects

COTS software components range from software development environments to operating systems, database management systems, and increasingly business applications. They come with a host of refined, feature rich, pre-configured and tested functionalities for most users of the application. These systems invariably offer rapid delivery of functionality to the end users, "shared" development costs with other customers and an opportunity for expanding capabilities and performance as improvements are realized in the marketplace. Typically, COTS systems have the following characteristics [Basili and Boehm, 2001]:

a. The buyer has no access to source code.
b. The vendor controls its development and
c. The software has a nontrivial installed base.

COTS products are often considered as the "silver bullet" in software engineering offering significant savings in procurement, design, development, testing and maintenance.

2. Hybrid Software projects

While "plug and play" i.e. COTS products has been fairly successful for hardware devices, it has not worked successfully for software. The challenge in a typical COTS product is to understand what are the functionalities currently offered in COTS and to what extent they meet the project requirements. COTS products are designed to meet the needs of a marketplace in general instead of satisfying the specific requirements of a particular organization. So these products generally need some "fine tuning" to adapt the functionality to the organization's needs. For instance, in an accounts payable application, the standard COTS product might come up with say 100 functions while the requirement analyst has elicited 25 functions from the users. The challenge is to understand if these 25 functions are a subset of the 100 pre-configured functions and if so to what extent it meets the needs of the users. So hybrid projects fundamentally adapt standard COTS product to the specific needs of the organization.

3. Bespoke Software Projects

While the COTS software allows the organization to focus on the business – not on software development, the downside of relying on COTS is that the product is not tailored precisely to the business process. In addition, non-development costs, such as licensing fees, are significant and more than half the features in COTS software products go unused [Basili and Boehm, 2001]. Table 1 below shows the advantages and disadvantages of COTS products [Boehm, 1999].

Table 1: Advantages and disadvantages of COTS

Advantages	Disadvantages
Immediately available; easier payback	Licensing, Intellectual property procurement delays
Avoids expensive development	Up-front license fees
Avoids expensive maintenance	Recurring maintenance fees
Predictable, confirmable license fees and performance	Reliability often unknown or inadequate; Scale difficult to change.
Rich functionality	Too-rich functionality compromises usability and performance.
Broadly used, mature technologies	Constraints on functionality and efficiency.
Frequent upgrades often in anticipation of organizational needs	No control over upgrades and maintenance.
Dedicated support organization	Dependence on vendor.
Hardware/software independence	Integration incompatibilities among vendors.
Tracks technology trends.	Synchronizing multiple-vendor upgrades.

Given the disadvantages in COTS products, bespoke software (commonly known as custom software) is a type of software that is developed for a particular organization for its specific needs. This software is inherently intuitive and exclusive and can give the company a unique competitive advantage in the market at expense of significant more costs and implementation time.

1.3 Software Project Execution

Software has become ubiquitous as it has penetrated almost every endeavor from business to leisure to healthcare and even our day-to-day needs. However it is disappointing to see that software projects today are often characterized by poor quality, schedule overruns and high costs in the private, the public, and the not-for-profit sectors across different industry segments around the globe.

The Standish Group's "CHAOS Summary 2009" report [Standish, 2009] says only 32% of all projects are delivered on time, within budget, with the required features and functions. About 44% were challenged as they were late, over budget, and/or with less than the required features and functions and 24% failed because these projects were cancelled prior to completion or never used [Standish, 2009]. The research also mentions that size does matter, and large projects are more likely to fail than small projects. Projects costing less than $750,000 in labor have a 71% chance they will be successful, while projects costing between $750,000 and $3 million have a 38% chance of being successful. Projects over $10 million only have a 2% chance of coming in on time and on budget. In addition some three quarters of all large systems are "operating failures" that do not function as intended or are not used at all. Moreover, the statistics from the Standish Group for the previous years as shown in the table 2 below [Standish, 2009] are also not very different from the 2009 statistics indicating that there are serious problems in executing software projects around the world.

Table 2: Standish Project Benchmarks

Year	Successful (%)	Challenged (%)	Failed (%)
1994	16	53	31
1998	26	46	28
2000	28	49	23
2004	29	53	18
2006	35	46	19
2009	32	44	24

Though the Standish Group's chaos report is challenged by some researchers [Sauer, Chris, Gemino, Andrew and Horner Reich, Blaize, 2007], other studies also confirm the high failure rate in IT projects. In 2004, a PriceWaterhouse Coopers study surveyed 10640 projects and revealed that only 2.5% of the projects achieve budget, schedule and scope targets [Dalcher, 2009]. Research by Dr John McManus and Dr Trevor Wood-Harper on IT projects carried out in the European Union between 1998to 2005 highlights that only one in eight information technology projects can be considered truly successful [McManus and Harper, 2007]. The dollar cost of IT failure worldwide is quantified at a staggering $6.2 trillion per year, or $500 billion each month [Sessions, 2010]. Essentially, all these statistics presented here on IT projects converge on:

1. Projects are more likely to be unsuccessful than successful.
2. Only about one third of the projects are likely to bring complete satisfaction to the stakeholders.
3. The larger the project the more likely the failure.

In spite of all these odds, organizations still execute software projects and expect them to be completed faster, cheaper, and with higher quality for the following key reasons:

1. The benefits of implementing a software project amongst others include improvements in employee productivity, and a higher degree of accuracy of information within the firm resulting

in better decision making and improving the organization's bottom line.

2. Many legacy systems can no longer meet organizations' business needs. For example, a recent study conducted by the Hackett Group found that four in five companies cannot accurately forecast midterm cash flow i.e. cash flow for the next two to three months. The main reason for this is the inability to automate the process from legacy applications into routine transactions, considering that cash flow is arguably the important measure for an organization [Simon, 2009].
3. Organizations can realize certain tax advantages by implementing a new system. Certain IT costs can be capitalized, impacting the overall net cost of the project. Capital expenditures (CAPEX) create future benefits for the organization. Although CAPEX alone is typically not the main driver for a new system, it is a nice side benefit and helps make the case for organizational change [Simon, 2009].

1.4 Software Project Status Reporting

There are many definitions of software project success and failure where each one is driven by different criteria. The classic definition of project success is completing the project on time, on budget, of high quality, with the expected features [Atkinson, 1999; PMI, 2008]. This definition however does not answer questions such as:

1. **When the project is considered success or failure? Is it immediately after the scheduled completion date or later when the ROI on the project is realized?**

   According to Darren Dalcher project outcomes should be a function of time scales, organization levels and goals. According to him a project can be seen as successful at four levels as shown in the table 3 below [Dalcher, 2009].

Table 3: Dalcher's definition of software project success

Level	Type	Critical Success Factors
1	Project Management Success	Efficiency and Performance
2	Project Success	Objectives, Benefits, Stakeholders
3	Business Success	Value Creation and Delivery
4	Future Potential	New Market, Skills, Opportunities

While levels 1 and 2 emphasize the delivery of the project, levels 3 and 4 encourage long term strategic thinking i.e. they look beyond single projects. Level 1 looks at the project management processes. This is not relevant for us as we are concerned with the project outcomes. As level 2 covers the **software development lifecycle** (SDLC) and p**roject management lifecycle** (PMLC) catering to the stakeholder's objectives, the focus of this research will be at level 2. In addition, technological and business changes are so rampant in software industry that measuring a project say two years from the project completion date for business success (level 3) and future potential (level 4) is farfetched. A classic example would be a standard feature turning into a defect a few years later because of changing business conditions or industry standards or even new government regulations.

2. **What are the levels of granularity in the success or failure and who defines them?**

To address this question Kurt Linberg proposed a new definition of software project success and failure as shown below in table 4 [Linberg, 1999]. This definition bridges the chasm between project team members and senior management perceptions of project success and failure by broadening the definition of success and failure.

Table 4: Linberg's definition of software project success and failure

	Completed Projects	Cancelled Projects
Failure	Does not meet customer expectations	Not learning anything that can be applied to future projects
Low Success	Below average cost, effort, and schedule performance compared to industry AND meeting quality expectations	Some learning can be applied to future projects
Success	Average cost, effort, and schedule performance compared to industry AND meeting quality expectations	Some learning can be applied to future projects and some artifacts can be used on future projects.
High Success	Better than average cost, effort, and schedule performance compared to industry AND meeting quality expectations	Substantial learning can be applied to future projects and a large number of artifacts used.
Exceptional Success	Meeting all quality, cost, effort and schedule expectations.	A Cancelled project cannot be considered an exceptional success.

The answers to the above 2 questions will help decision makers to understand the definition of project success or failure better and take corrective actions such as fast tracking, crashing, adding additional resources (people, tools, time, money etc.), scope reduction, extending timelines to name a few.

1.5 The Need for a Software Project Measurement Framework

We best manage what we can measure. According to Barry Boehm, "Software measurement can be loosely defined as the process of defining, collecting and analyzing data on the software development process and its products in order to understand and control the process and its products, and to supply meaningful information to improve that process and its products" [Goodman, 2004]. Some of the important benefits of metrics to the software project and the organization are [Kan, 2003; Kaner and Bond, 2004]:

1. Metrics provide **objective information** throughout the software organization across the project lifecycle. Metrics help to answer important questions such as "Where are we now?", "Is the project on schedule?" or "Is the software ready to be delivered to the user?" etc.
2. Measurement helps to **manage project issues** i.e. identify, prioritize, track and communicate at all levels within the organization.
3. Metrics **promote teamwork and improve team morale** by linking efforts of individual team members with the overall objectives of the project.
4. Metrics facilitate a **proactive management strategy**. Potential problems are objectively identified and existing problems can be better evaluated and prioritized. Metrics foster the early discovery and correction of technical and management problems that can be more difficult or costly to resolve later.
5. Metrics provide an effective **rationale for selecting the best alternatives** when software projects are usually constrained by ambiguity and complexity.
6. Metrics on effort, schedule, scope and defects for a number of projects (programs and portfolio) provide opportunities to **analyze** trends, ultimately resulting in more accurate predictions for **future projects**.

According to Watts S Humphrey, based on the business needs there are four types of software metrics [Humphrey, 1999]:

1. **Understand:** Understand type metrics help us to know about the software processes, products and services.
2. **Evaluate:** Evaluate type metrics are used in the decision-making process to study products, processes or services in order to establish baselines and to determine if established standards, goals and acceptance criteria are being met.
3. **Control:** Control type metrics are used to monitor our software processes, products and services and identify areas where corrective or management action is required.
4. **Predict:** Predict type metrics are used to estimate the values of base or derived measures in the future.

Essentially each of these four types of metrics should function in any one or more of the three ways:

1. **Informational** - Informational metrics are used to track progress, results, value, and performance to budget, schedule and quality; typically the triple constraints of the project.
2. **Diagnostic** - Diagnostic metrics are used to expose problem areas.
3. **Motivational**- Motivational metrics are used to influence the behavior of team members or stakeholders in the project.

1.6 Key Challenges in Deriving the Right Measurement Framework

While it would appear that poor success rate in software projects is primarily caused by the inherent complexity in projects, one of the approaches to deliver a successful software project is formulating a measurement framework so that suitable corrective actions in the project can be taken at the right time. Measurement provides quantitative descriptions of key processes and products to have a precise, predictable, and repeatable control over the software project [Pfleeger, S.L, Jefferey, R, Curtis, B and Kitchenham, B, 2002]. Successful projects are well managed and controlled and the control mechanism is measurement [Vierimaa, M, Tihinen, M, Kurvinen, T, 2001].

Given that measurement is the basis for detecting deviations from acceptable performance, an organization's method of monitoring project status is generally seen as a reflection of the project management maturity. In fact, sound measurement practices are integral to basic management activities such as project planning, monitoring, and control [Rad and Levin, 2005]. Dr Harold Kerzner's 16 points to project management maturity includes measuring project progress periodically [Kerzner, 2003]. The SEI CMMI (Software Engineering Institute - Capability Maturity Model Integration)-Level 5 certification advocates applying statistical techniques for repeatable project performance. According to Roger Pressman, the software project management activities should include measurement [Pressman, 2004]. Karl Wigers includes "Track project status openly and honestly" as one of the top 21 secrets of successful software Project Management [Weigers, 2002]. John Reel includes "Track Progress" as one of the five essential factors for successfully managing a software project [Reel, 1999]. Larry Putnam says that software projects can be managed for success with progress indicators applied through metrics based management [Putnam, 2002].

Though metrics give us the direction to better manage a project, it is often the most complicated project management process and is usually ignored or handled poorly. Hence projects never see the light of the day as stakeholders including project managers do not have all the facts and figures on how the project is performing at various points in time to take appropriate corrective measures. According to Watts Humphrey, management visibility is a problem in software projects as the project manager generally cannot tell where the project stands [Humphrey, 2005]. He advocates - to manage software projects, one must know the status on where the project stands, how rapidly the work is being done and the quality of the products being produced for project success [Humphrey, 2005].

Though there are many reasons for not having the right metrics, three important reasons stand out which are explained below.

1.6.1 Varied Interests of Project Stakeholders.

Project stakeholders are individuals and organizations who are actively involved in the project and exert their influence over the project's objectives and outcomes. The stakeholders generally have varied level of interest, involvement and influence on the project and their goals vary and conflict reflecting different political and organizational interests.

While the project sponsor(s) and management in most projects are interested in schedule, cost, Total Cost of Ownership (TCO) and Return on Investment (ROI), the software users are looking at the quality of the software product for the agreed scope. In fact, when Dr Dobb's journal ran a survey in August 2007, 87% of the respondents indicated that high quality software is more important than being on budget and time [Ambler, 2008]. In addition to the conflicting goals, in most projects the expectations of the stakeholders are implicit and subjective.

Hence derivation of measures is extremely complex as it generally combines the different levels of stakeholder expectations over different timescales for different types of results as shown below in table 5 [Turner, Zolin and Remington, 2009].

Table 5: Expectations of the Project Stakeholders

Results	**Project Output**	**Project Outcome**	**Impact**
Time Scale	**End of Project**	**Plus Months**	**Plus Years**
Stakeholder			
Users	Features	Usability	New technology
	Performance	Convenience	New capability
	Documentation	Reliability and Availability	New competence
	Training	Maintainability	New class
			Job security

Project sponsor and Management	Features	Performance	Future projects
	Performance	Benefits and Reputation	New technology
	Time and cost	Relationships	New capability
	Completed work	Investor Loyalty	New competence and class
Project Manager and Project team	Time and cost	Reputation	Job security
	Performance	Relationship	Future projects
	Learning	Repeat business	New Technology
	Camaraderie		New competence
	Retention		
	Well being		
Suppliers	Time and cost	Performance	Future business
	Performance	Reputation	New technology
	Profit	Relationship	New competence
	Client appreciation	Repeat business	Whole life social cost-benefit ratio
	Safety record, Environmental impact	Environmental impact	
	Completed work	Social costs and benefits	

1.6.2 Challenges due to the Inherent Nature of Software projects

By their nature software projects are generally dynamic; they tend to grow, change, and behave in ways that one cannot always predict. For example, one cannot build a software project in half the time with twice the people, although ratio math would allow this. Though software can be delivered under certain controlled environment conditions and very specific functional structures, it is usually governed by constraints which are multidimensional and abstract in nature. In this backdrop, the inherent nature of software projects is based on three main elements [Remington and Oollack, 2008; Brooks, 1995].

1. Complexity

Nearly all software projects exhibit some degree of complexity. A complex system is an entity which is coherent in some recognizable way but whose elements, interactions and dynamics generate structures allowing surprise and novelty which cannot be defined in advance [Batty and Torrens, 2005]. Edward W. Felten, a computer science professor at Princeton University, quoted: "A corporate computer system is one of the most complex things that humans have ever built". Fred Brooks, in his classic book "The Mythical Man-Month", states that "Software entities are more complex for their size than perhaps any other human construct. Many of the classical problems of developing software products derive from this essential complexity [Brooks, 1995, pp 182]". According to software architect Roger Sessions, the primary cause of software project failures is complexity [Sessions, 2010].

Understanding the types of project complexity is important because the source of project complexity will influence the project lifecycle, duration of project phases within the lifecycle, the governance structure, selection of key resources, scheduling and budgetary methods and ways of identifying and managing risks. According to Remington and Oollack, there are four types of project complexity and any project will exhibit one or more types of complexity [Remington and Oollack, 2008]. They are:

1. Structural Complexity
2. Technical Complexity

3. Directional Complexity
4. Temporal Complexity

2. Structural Complexity

This kind of complexity is normally found in large projects due to the difficulty in managing and keeping track of the huge number of different interconnected tasks and activities. To manage these projects, outcomes are decomposed into many small deliverables which can be managed as discrete units. The underlying assumption is that the individual units, when delivered, will come together to make the required whole. Examples include testing complexity which is dependent on design and coding.

3. **Technical/System complexity**

This type of complexity is found in projects which have technical or design problems associated with products that has never been produced before. This is also known as system complexity and encompasses three dimensions [Xia and Lee, 2005]:

i. Variety. It is the multiplicity of project elements such as number of sites or applications affected by the system implementation.

ii. Variability. This refers to changes in project elements such as changes in project goal and scope.

iii. Integration. This taps into coordination of various project elements.

3. Directional complexity

Directional complexity is found in projects which are characterized by unshared goals and goal paths, unclear meanings and hidden agendas. This kind of complexity stems from ambiguity related to multiple potential interpretations of goals and objectives.

4. Temporal complexity

These projects are characterized by shifting environmental and strategic directions which are generally outside the direct control of the project team. This kind of complexity stems from uncertainty regarding future constraints, the expectation of change and possibly even concern regarding the future existence of the system. Temporal complexity can be found in projects which are subjected to unanticipated environmental impacts significant enough to seriously destabilize the project, such as the development of new technologies, competition, government regulations etc.

5. Uncertainty

Closely correlated with complexity is uncertainty. Today a vast majority of software projects are still performed in a "deterministic" manner where the tasks and resources are assigned and executed within well defined timeframes following the traditional waterfall model. At the same time practical situations necessitates corrective actions when the right resources are not always available on demand, additional tasks pop up in the middle of the execution, scope creep occurs and requirements changes, to name a few. Uncertainty is the existence of more than one possibility (positive or negative) where the outcome is not known because of information inadequacy. It cannot be eliminated completely but there will always be some degree of uncertainty at different stages in the project.

Though uncertainty is not desirable it can sometimes be useful in projects as it might throw some opportunities not just for the project but also for the organization [Ward, Stephen and Chapman, Chris, 2003]. It is particularity high during the early stages of the project and leading to increased risks. Uncertainty leads to assumptions and it increases with the number of people in the project. These assumptions can be on business processes, project management practices, people, dependencies on other streams, requirements, design, development, testing, deployment etc. But when many of these assumptions are not clearly apparent or discovered, conflicting assumptions can exist for quite some time before they manifest themselves as problems. These problems can potentially lead to re-planning, redesign, and rework affecting project schedule and budgets.

6. Invisibility

The biggest difference between software and other kinds of engineering projects is that software is not physical or visible. According to Frederick Brooks, "such essential complexity is unique to software development because software is invisible and unvisualizable, and is subject to conformity and continuous changes" [Brooks, 1995, pp 183]. Due to the lack of visualization of the software product, customer i.e. the user cannot have any underlying sense of what is attainable and may ask for functions that are impossible to deliver. Their inability to visualize the boundaries creates indifference to what is possible and what is not and encourages the software users to change their minds on requirements more frequently [Gross, 2006]. In addition, invisibility has the danger of creating a misperception that anything and everything is possible in software making both customers and developers susceptible to forgetting the limitations of software.

All the three attributes discussed above – complexity, uncertainty, and invisibility are inherent, co-exist and affect each other as shown in the figure 1 below. The three attributes should be managed as they cannot be eliminated completely; though the level of intensity varies depending on the phase of the project lifecycle.

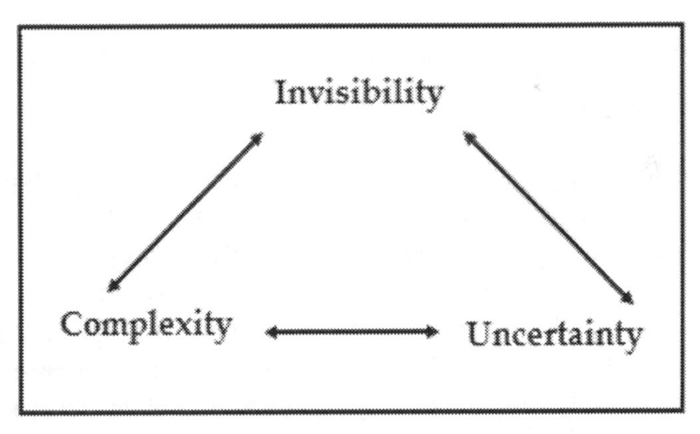

Figure 1: Software Project Attributes

Coupled with these three unique characteristics, IT is a much younger branch compared to other engineering disciplines, and the Project

Management Maturity (PMM) is also much lower compared to other engineering branches as shown in table 6 below [Schwalbe, 2005].

Table 6: Project Management Maturity Comparison

1 = Lowest Maturity Rating, 5 - Highest Maturity Rating				
Knowledge Area	Engineering/ Construction	Telecommunications	Information Systems	Hi-Tech Manufacturing
Scope	3.52	3.45	3.25	3.37
Time	3.55	3.41	3.03	3.50
Cost	3.74	3.22	3.20	3.97
Quality	2.91	3.22	2.88	3.26
Human Resources	3.18	3.20	2.93	3.18
Communications	3.53	3.53	3.21	3.48
Risk	2.93	2.87	2.75	2.76
Procurement	3.33	3.01	2.91	3.33
Total	**26.69**	**25.91**	**24.16**	**26.85**

1.6.3 Misalignment between Measurement researchers and Industry practitioners.

According to Pfleeger et al, there is a discontinuity in measurement knowledge between researchers and practitioners. Software metrics researchers are usually more interested in validating theoretical concepts and conclusions, while practitioners usually want "few, quick and useful metrics" that have been empirically tested [Pfleeger, Ross, Curtis and Kitchenham, 1997]. Though there are a large number of measures, only a few have enjoyed any widespread use or acceptance. For example, according to SEI there are 31 measures on product complexity [Mills, 1988]. But do we have time in a project to use them all to measure complexity? Even worse, over 300 measures are identified for software development alone [Far, 2008]. Even in the case of widely studied metrics such as

LOC, Halstead's metrics, and McCabe's cyclomatic complexity, it is not universally agreed what they measure [Mills, 1988].

It would appear that academic software metrics research has failed almost totally in terms of industrial penetration. According to Fenton, academic research is irrelevant for the industry at two levels [Fenton, 2006]:

- **Irrelevance in scope.**
  Much academic work has focused on metrics which can only ever be applied or computed for small programs, whereas all the reasonable objectives for applying metrics are relevant primarily for large systems. Irrelevance in scope also applies to the many academic models which rely on parameters which could never be measured or implemented in practice.

- **Irrelevance in content.**
  While the pressing industrial need is for metrics that are relevant for process improvement, much academic work has concentrated on detailed code metrics. In many cases these aim to measure properties that are of little practical interest to the practitioners.

The software industry has also contributed its share for the failure of software metrics as well [Fenton, 2006]:

- **Much of the industrial metrics are poorly motivated.**
  In most cases the decision by a company to put in place some kind of a metrics program is done when things are bad or in order to satisfy some external assessment body. For example, in the US the single biggest trigger for industrial metrics activity has been the CMMI [Humphrey, 1989]. Convincing success stories describing the long-term payback of metrics are almost non-existent as metrics are generally considered an overhead on the current software projects at 4-8% [Hall and Fenton 1997]. So when deadlines are tight and slipping it is inevitable that the metrics activity will be one of the first things to suffer [Fenton, 2006].

- **Much of the industrial metrics activity is poorly executed.** In addition software industrial metrics activity is poorly executed ignoring well-known guidelines of best practice data-collection and analysis [Fenton, 2006]. Industry averages show that only 20 percent of measurement programs implemented survive past the two-year mark [Dekkers and McQuaid, 2002].

However over the years, much research work has even focused on meta-level activities such as:

1. The work of Robert Grady which is an extensive experience report of the software metrics program in HP [Grady, 1992].
2. The work on GQM (Goal-Question Metric) [Basili, 1994].
3. The use of metrics in empirical software engineering (ESE) as pioneered by Basili where success in research is judged by the acceptance of empirical results and the ability to repeat experiments to independently validate results.
4. Work on theoretical underpinnings of software metrics using measurement theory as basis for software metrics [Briand, El Emam and Morasca, 1996; Fenton, 1991, Zuse, 1991].

To summarize, these three main reasons (i.e. Varied stakeholder interests, inherent challenges in software projects, and misalignment between researchers and practitioners) have resulted in firstly, lack of maturity in software measurement and secondly, no standardization of software measures. This has ultimately resulted in the poor application of measurement frameworks in software projects with most decisions in software projects and programs being done intuitively without understanding the complete picture of the project.

1.7 The Research Question

The outcome of the literature studies is the following research question:

How can we have a *generic* and *objective measurement framework* for software projects?

This research question has three key words/phrases:

1. **Measurement framework**
   - Which measurement framework to apply in software projects given the vast number of measurement frameworks such as GQM, PSM, BSC etc available?

2. **Generic**
   - How to derive a generic "One-size-fits-all" measurement model given that every software project is unique and measurement/project status is context sensitive?
   - If the validation of the measurement framework helps in generalization (and hence reliability) what are the validation criteria?

3. **Objective**
   - How can we quantify the critical success factors (CSFs) of the project stakeholders which are mostly abstract and implicit?

In other words, the main research question can be framed as - **how can software measurement benefit industry practitioners?** While deriving a stakeholder driven generic and objective measurement framework is important, it should also be validated – theoretically and empirically. This is in concurrence with the research findings of Norman Fenton who proposed [Fenton, 2006]:

1. A sound conceptual **theoretical validation** where the measurement framework is structurally sound in accordance to the measurement theory.

2. Statistically significant **empirical validation**. This is needed for conformance of the measures and metrics to the real world and to complement the theoretical validation. This

is needed because most metrics have been defined, applied and tested in a very limited environment and have been unsuccessful in predicting events. Subsequent attempts to test or use the metrics in other environments have yielded very different results owing to lack of clear definitions and testable hypothesis.

1.8 What is not in Scope for this Research

While the research question outlines the scope of research, below are some of the items which are outside the scope of the proposed research.

- New measures will not be developed. There are many proven measures available that have been extensively researched and no effort will be consciously made to identify new measures.

- Subjective project measures such as customer satisfaction (CSAT), team morale etc are not in scope. When working with a subjective measure, the person making the measurement takes decisions by making some sort of judgment. The subjective measure depends on the object and the viewpoint from which they are taken. The value given by the measure can be different if the object is measured again [Wohlin et al, 2000].

- The tools including the methodology (For example requirements engineering, effort estimation etc) for capturing the data for the measures are not in scope. Most development editors have inbuilt tools and in addition there are numerous freeware tools available to compute product metrics. For the computation of project and process metrics, there are tools such as MS project, Primavera etc which are available. Implementing the measurement framework involves many dimensions and should be tailored on the basis of project situations. Hence this research thesis does not attempt to come up with tools and a methodology for measurement.

- Benchmarking of the measures and the results. The data on the measures will not be used to compare projects as every project success criteria depends on the specific thresholds, variances and control limits set by the stakeholders.

- Decision making rules from the measures. The objective of this research is to derive a measurement framework to provide information for decision making for project stakeholders. The decision makers can use this measurement framework to complement their experience and intuition and take appropriate corrective action.

1.9 Conclusion

Today speed of development (or time to market), reduced cost and high quality are the priorities in software delivery. Hence management by projects is being recognized as one of the most effective ways of managing an organization's needs and responding to change combining software development and project management practices. Projects integrate activities within an organization including strategies, priorities, and resource pools providing order and control over the activities performed by individuals. However given the poor track record of software project success, improvement is desired in the project delivery. According to Peter Drucker, "What is measured improves" [Drucker, 2001]. To manage projects and to take appropriate corrective actions for improvement, one has to know where the project stands, and for taking appropriate corrective actions we need a generic, objective and validated software measurement framework.

This book i.e. the thesis is organized as follows. Chapter 2 covers the research methodology including the validation criteria to be applied on the measurement framework. In chapter 3, based on the Goal-Question-Metric (GQM) model, a generic and objective GQM measurement framework using six steps addressing the concerns/goals of the project stakeholders is formulated. In chapter 4, the measures derived using the GQM framework is discussed in detail. Chapter 5 covers the theoretical validation criteria from measurement theory proposed by Kitchenham, Fenton and Pfleeger.

At the measurement framework level, the ten questions of Cem Kaner and Walter Bond are applied for compliance and to serve as a bridge between theoretical and empirical validation. Though theoretical validation using measurement theory is getting a great deal of attention from researchers, industry practitioners still rely on empirical evidence of a measure's utility. Hence this research is empirically validated in Chapter 6 using criteria (six proposed by Schneidewind and two proposed by Kitchenham, Fenton and Pfleeger) with a survey and case studies (controlled and uncontrolled). Chapter 7 concludes this thesis with the conditions of implementing this framework and directions for future research.

Chapter 2: Research Design and Methodology

2.1 Introduction

Research design refers to the structure of an enquiry to ensure that the evidence obtained answers the research question as unambiguously as possible. It also provides the blueprint for the collection, measurement and analysis of data. Rigorous attention to design details encourages an investigator to focus the research method on the research question, thereby bringing precision and clarity to the study [McGaghie, William C, Bordage, Georges, Crandall, Sonia and Pangaro, Louis, 2001]. Specifically research design should provide:

1. An activity and time based plan to answer research problems such as [Philliber, Schwab and Sloss, 1980]:

   - What questions to study?
   - What data are relevant?
   - What data to collect?
   - How to analyze the results?

2. A road map i.e. methodology for the researcher to focus on the research problem(s) for an orderly approach to the collection, analysis, and interpretation of data that addresses the research problem(s).

The research design can be descriptive or explanatory and should typically encompass five dimensions [McGaghie et al, 2001].

1. **Appropriateness of the design**. The research design should be clearly defined to permit the study to be replicated.

2. **Internal validity**. This focuses on a addressing the sources of bias or confounding variables, including selection bias, intervention bias, measurement bias and many more which impact the integrity in rigorously addressing the research question rigorously.

3. **External validity**. Are the research results generalizable to subjects, settings, and conditions beyond the research situation?

4. **Unexpected outcomes**. Are allowances made for expression of surprise results the researcher did not consider or could not anticipate? Any research design too rigid to accommodate the unexpected may not properly reflect real-world conditions or may stifle the expression of the true phenomenon studied.

5. **Plausibility**. Common flaws in research include failure to randomize correctly, small sample sizes resulting in low statistical power, brief or weak experimental interventions, and missing or inappropriate comparison (control) groups. Often there are tradeoffs in research between theory and pragmatics, precision and richness, elegance and application and the research design should be attentive to such compromises.

In this scenario, the research methodology must as far as possible should be controlled, rigorous, systematic, valid and verifiable, empirical and critical. The main steps in the research methodology for the research thesis are as shown in the figure 2 below.

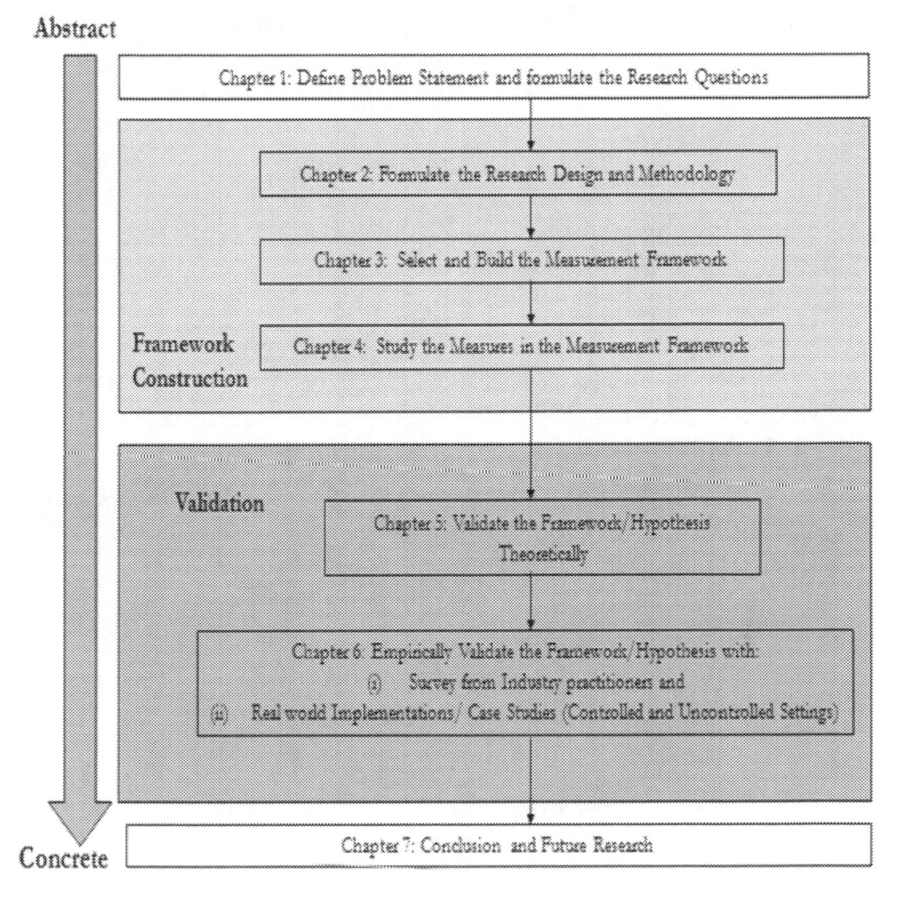

Figure 2: Steps in the Research methodology

Each chapter in general follows the "ETVX" principles where 'E' is the entry criteria which must be accomplished before a set of tasks can be performed, 'T' is the set of tasks to be performed, 'V' stands for the verification and validation process to ensure that the right tasks are performed, and 'X' stands for the exit criteria or the outputs of the tasks. The outcome of the research process is to target a validated, objective and a generic stakeholder driven measurement framework for understanding, predicting and controlling of software projects.

2.2 Evaluation of Measurement Framework

The most important criterion for the evaluation of any measurement framework is the adequacy of measurement. This adequacy can be evaluated in terms of:

1. Reliability of measurement
2. Validity of measurement.

Reliability estimates the consistency of the measurement, or more simply the degree to which the measuring instrument measures the same way each time it is used under the same conditions with the same subjects. Validity, on the other hand, involves the degree of accuracy the measurement is done. Together reliability and validity form the core of what is accepted as scientific proof which is as shown in the table 7 below [Kothari, 2001].

Table 7: Reliability and Validity

	Valid	**Not Valid**
Reliable	You are measuring what you think you are measuring, and doing it reliably	You have a reliable measure of something, but it is biased or it is not measuring what you thought it should.
Not Reliable	The average measurement is right on, but each individual measurement has so much error in it that it is unusable by itself	Each time you measure you get something different, and the average measurement is way off the target.

More simply, validity is the accuracy while reliability is the repeatability of the measurement. Figure 3 below portrays the same idea [Kothari, 2001].

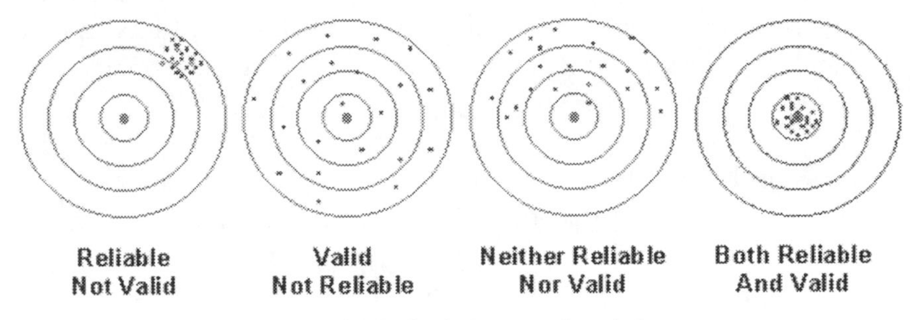

Figure 3: Reliability and Validity

2.2.1 Reliability

Reliability is essentially the consistency of the measurement indicating the extent to which the measured value is error free i.e. any departure of the result from the "true" value i.e.

Measured value = True Value + Systematic Error (i.e. Bias) + Random Error (precision)

Whenever measurement procedures involve human intervention, we have to question whether the results are reliable or consistent. So to determine whether two data sets are consistent in their observations, we have to determine how reliable the data sets are. The key point in reliability is that it is not measured, it is estimated and the accuracy of the estimation can be improved by having a good sample size.There are two ways reliability is estimated:

1. Test/Retest reliability test
2. Internal Consistency test.

The primary difference between the two is that test/retest involves two administrations of the measurement instrument, whereas the internal consistency method involves only one administration of that instrument.

A. Test-Retest

Test-Retest is one of the simplest ways of testing the reliability of an instrument over time and assumes that there will be no change in the quality or construct being measured. The idea behind Test-Retest is that one should get the same score on test 1 and test 2. The three main steps in this method are:

1. Run the tests test 1 and test 2 at two separate times for each subject ensuring that there is no change in the underlying condition (or trait you are trying to measure) between test # 1 and test # 2.
2. Compute the correlation between the two tests i.e. test 1 and test 2.
3. If the correlations between two tests is high (i.e. 0.7 or higher) then there is a good test-retest reliability.

The disadvantage with Test-retest reliability test is the potential for a carryover effect between testing. The first testing may influence the second testing resulting in the contamination of scores. This effect can however be reduced by having a sufficient time gap between the two tests.

B. Internal Consistency Test

Internal Consistency reliability test is performed using only one test administration and thus avoids the problem associated with repeated testing. Measuring internal consistency involves measuring two different versions of the same item within the same test.

There are three main techniques for measuring the internal consistency based on the computational complexity, robustness, scope, and speed of the test.

1. **Split-Halves Test**

   The split halves test for internal consistency reliability is the easiest type, and involves dividing a test into two halves randomly. The results from both halves are statistically analyzed, and if there is weak correlation (less than 0.5) between the two, then there is a reliability problem with the test. Split-halves testing is a popular way to measure reliability, because of its simplicity and speed [Cooper and Schindler, 2000].

2. **Kudar-Richardson (KR) test**

   The Kudar Richardson (KR) test works out the average correlation for all the possible split half combinations in a test. For example, if we have four items we will have six different correlations i.e. ${}^4C_2$. The average correlation is simply the average or mean of all these correlations. For example, if the individual correlations for the four different tests are 0.86, 0.90, 0.90 and 0.94, the average correlation will be 0.90 [Allen and Yen, 2001].

3. **Cronbach's Alpha Test**

   Cronbach's Alpha can be viewed as an extension of the KR test. Cronbach's Alpha is mathematically equivalent to the average of all possible split-half estimates. In this test we compute one split-half reliability and then randomly divide the items into another set of split halves and re-compute it again. We keep doing this until we have computed all possible split half estimates of reliability.

The test conducted is as shown in figure 4 for one set of sample measured after one administration. The values 1, 2, 3 etc are different DRE readings of the subjects in the sample. Value 1 could be the reading given by person 1 using a measurement instrument and process, Value 2 could be a reading given by person 2 using the measurement instrument and process etc. Essentially, Cronbach's alpha splits all the questions on the measurement instrument in every possible way and computes correlation values for all of them and generate one number known as Cronbach's alpha. If this value is greater than 0.7, the data set is accepted for reliability [Allen and Yen, 2001].

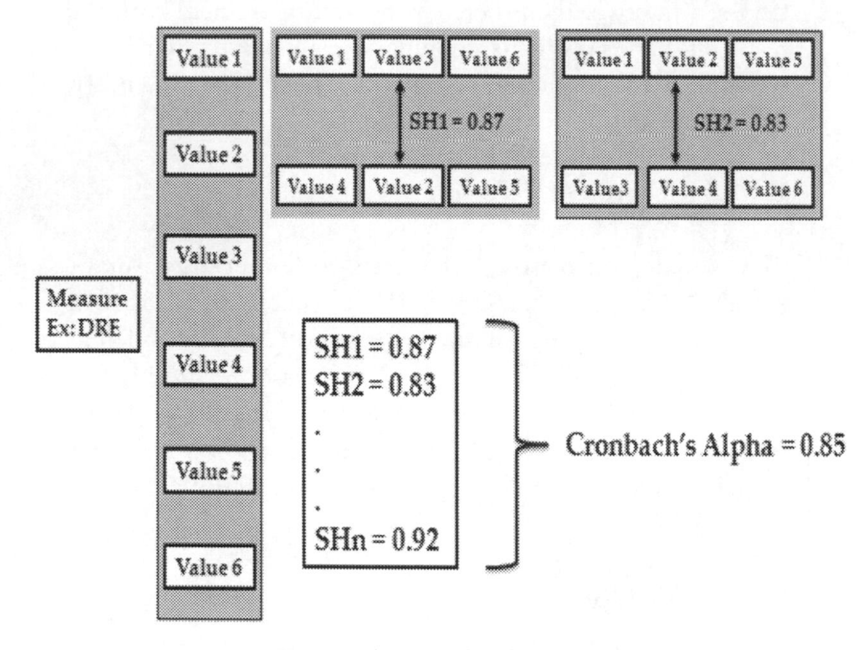

Figure 4: Cronbach's Alpha

Of the three tests, Cronbach's Alpha is preferred over Split-halves test and KR Test. Cronbach's Alpha is known to be a less conservative estimate of reliability than test-retest and the other two types of IC tests i.e. Split-Halves and KR test [Allen and Yen, 2001].

2.2.2 Validity

Validity is concerned with whether we are measuring what we intend to measure or how the observations are influenced by the circumstances. It generally refers to the extent to which a concept, conclusion or measurement is well-founded and corresponds accurately to the real world. Validity is determined by a body of research that demonstrates the relationship between the test and the behavior it is intended to measure. In this backdrop, validity is classified into four types with each type of validity addressing one specific question [Trochim, 2002].

1. **Conclusion Validity**: Conclusion validity is concerned with whether there is any kind of relationship between the variables (dependent and independent) being studied.

2. **Internal Validity**: Assuming that there is a relationship between the dependent and independent variables, internal validity probes whether the relationship is a causal one?

3. **Construct Validity**: Assuming that there is a causal relationship between the dependent and independent variables, can we claim that the measure reflected well the idea of the construct of the measure? Construct validity asks did we operationalize well the ideas of the cause and the effect?

4. **External Validity**: Assuming a causal relationship in this study, can we generalize this result to other situations?

These four validity types build on one another where each validity type presupposes an affirmative answer to the previous one. The figure 5 below shows the idea of cumulativeness, along with the key question for each validity type [Trochim, 2002].

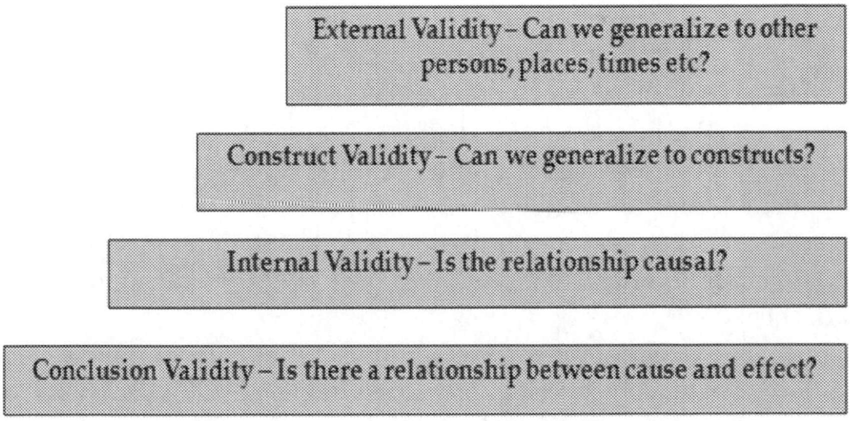

External Validity – Can we generalize to other persons, places, times etc?

Construct Validity – Can we generalize to constructs?

Internal Validity – Is the relationship causal?

Conclusion Validity – Is there a relationship between cause and effect?

Figure 5: Validity Hierarchy

2.3 Validation – Identifying the Criteria for Theoretical and Empirical Validation

The key component of the research question is validation of the measurement framework. Generally validation is not a binary trait, but rather a degree of confidence; greater the evidence the more valid is the measure.

As discussed while framing the research question in section 1.7, the measurement framework must be validated theoretically and empirically. Theoretical validation tells us if a measure is valid with respect to certain defined criteria from measurement theory, while empirical validation is to provide corroborating evidence of validity or invalidity from real world instances. Though theoretical validation with measurement theory is getting a great deal of attention from researchers, industry practitioners still rely on empirical evidence of a measure's utility. However neither kind of validation is sufficient by itself and both theoretical and empirical validation must be carried out to avoid the problem of defining a theoretically sound measure which is otherwise useless in its practical application [Braind et al, 1995].

However, what constitutes the software metrics validation criteria has been intensely debate for almost half a century and researchers have not yet reached a consensus on this system of rules because no researcher has provided a "proper discussion of relationships among different approaches to metrics validation" [Kitchenham et al, 1995]. Instead, the metric researchers have often been proposing their own, specialized means of validation.

Hence, in the absence of a generalized and accepted validation approach, the process followed by Meneely et al is adopted to find the criteria for both theoretical and empirical validation. Meneely et al carried out a systematic literature review of 2,288 peer-reviewed papers pertaining to software metrics validation and ultimately extracted 47 unique validation criteria from 20 papers as shown in table 8 [Meneely, Smith and Williams, 2010].

Table 8: List of all 47 validation criteria

1. A priori validity	25. Monotonicity
2. Actionability	26. Metric Reliability
3. Appropriate Continuity	27. Non-collinearity
4. Appropriate Granularity	28. Non-exploitability
5. Association	29. Non-uniformity
6. Attribute validity	30. Notation validity
7. Causal model validity	31. Permutation validity
8. Causal relationship validity	32. Predictability
9. Content validity	33. Prediction system validity
10. Construct validity	34. Process or Product Relevance
11. Constructiveness	35. Protocol validity
12. Definition validity	36. Rank Consistency
13. Discriminative power	37. Renaming insensitivity
14. Dimensional consistency	38. Repeatability
15. Economic productivity	39. Representation condition
16. Empirical validity	40. Scale validity
17. External validity	41. Stability
18. Factor independence	42. Theoretical validity
19. Improvement validity	43. Trackability
20. Instrument validity	44. Transformation invariance
21. Increasing growth validity	45. Underlying theory validity
22. Interaction sensitivity	46. Unit validity
23. Internal consistency	47. Usability
24. Internal validity	

2.4 Refining the Criteria List – First Iteration

It is practically not viable for a measure to satisfy all the 47 criteria as they are proposed by different authors from different backgrounds and contexts [Meneely et al, 2010]. The list of 47 validation criteria is meant to be reference. In addition, the 20 papers involved significant cross-referencing, discussion, and disagreement as shown in figure 6 indicating significant duplication of the validation criteria [Meneely et al, 2010].

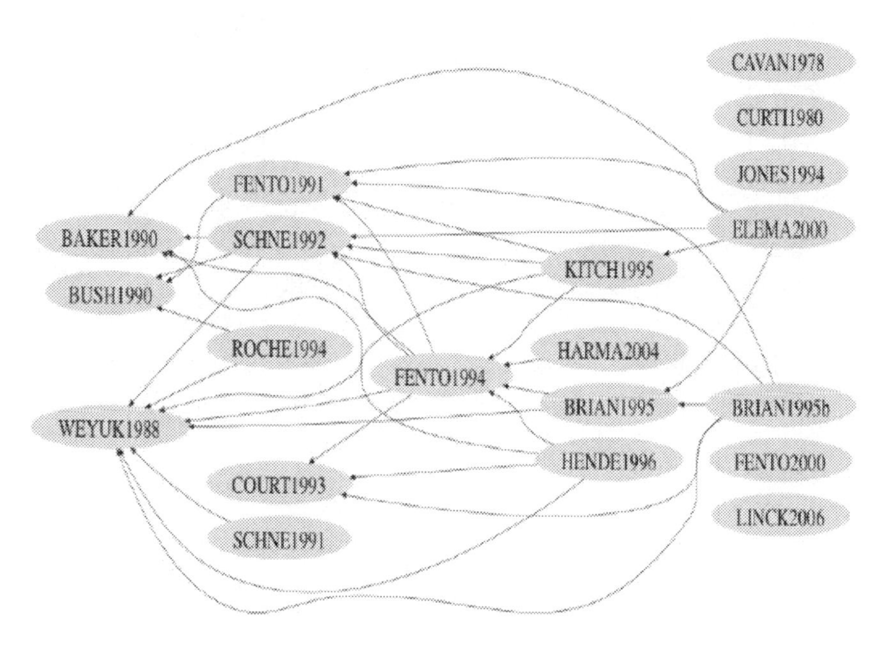

Figure 6: Citation Network

Hence for more practical viability, it was decided to bring the list down to two or three top authors/papers so that the measurement framework can be validated against their work. The criterion for being the top author amongst these 20 papers (authors) is based on the number of citations in Google scholar. Hence these 20 papers were further queried in Google scholar to find the number of citations for each one of them. The outcome as on 1st February 2011 is as shown in the figure 7 below.

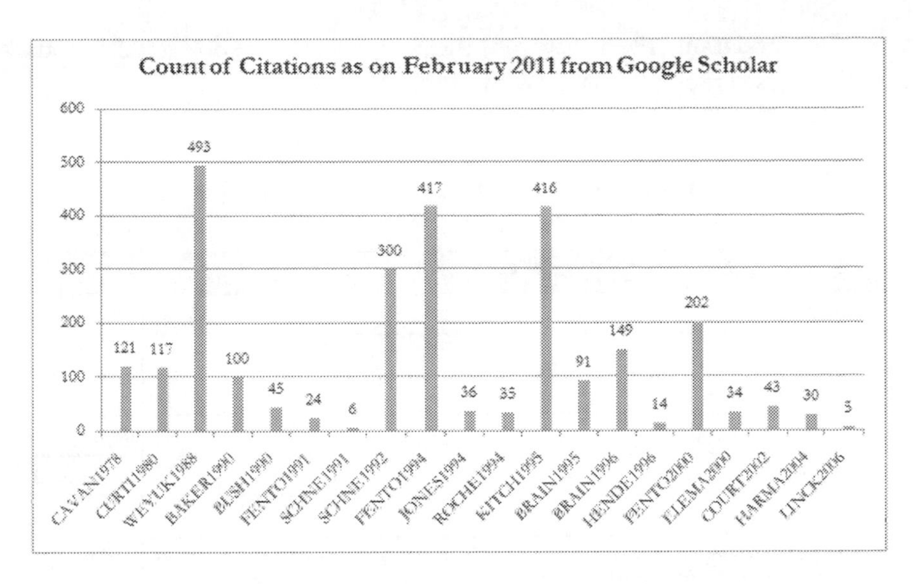

Figure 7: Citations for Top Papers

From this list of 20 papers, any paper which was published in the last 20 years and has citations of over 150 was selected. This provided concurrency and relevancy of the research papers and this in turn provided a list of top 6 papers frequently cited papers as shown in table 9 below.

Table 9: Top Six Papers

SL #	Author_Year	Name of the Paper	# of Citations as on February 1st 2011
1	WEYUK1988	Evaluating Software Complexity Measures	493
2	FENTO1994	Software Measurement: A Necessary Scientific Basis	417
3	KITCH1995	Towards a Framework for Software Measurement Validation	416
4	SCHNE1992	Methodology for Validating Software Metrics	300
5	FENTO2000	Software Metrics: Roadmap	202
6	BRAIN1996	On the application of Measurement Theory in Software Engineering	149

The validation criteria list from these 6 papers came down to 28 from 47 and these criteria are as shown in table 10 below.

Table 10: List of 28 validation criteria

SL #	Criteria	WEYUK1988	SCHNE1992	FENTO1994	KITCH1995	BRAIN1996	FENTO2000
1	A priori validity			Y		Y	
2	Actionability						Y
3	Appropriate Continuity				Y		
4	Appropriate Granularity	Y			Y		
5	Association		Y	Y			
6	Attribute Validity				Y		
7	Causal Model Validity						Y
8	Discriminative Power		Y				
9	Dimensional Consistency				Y		
10	External Validity			Y		Y	
11	Instrument Validity				Y		
12	Increasing growth Validity	Y					
13	Interaction Sensitivity	Y					
14	Monotonicity	Y					
15	Non-Uniformity	Y					
16	Permutation Validity	Y					
17	Predictability		Y	Y			
18	Process/Product Relevance		Y				
19	Protocol Validity				Y		
20	Rank Consistency		Y				
21	Renaming Insensitivity	Y					
22	Repeatability		Y				
23	Representation Condition			Y	Y		
24	Scale Validity			Y	Y		
25	Theoretical Validity			Y		Y	
26	Trackability		Y				
27	Underlying Theory Validity				Y		
28	Unit Validity			Y	Y		

2.5 Refining the Criteria List – Second Iteration

To further prune the list, each of the papers was studied in detail. Though, Weyuker's paper was in the list above i.e. WEYUK1998, it was decided to not include Weyuker's validation criteria for the following three reasons:

1. Weyuker's seven criteria identified in table 10 are specifically defined for code complexity measures which are relevant mainly for the software product. The validation of the proposed measurement framework is on a generic core set of measures in a software project encompassing product, process and resources.
2. Her criteria have been intensely contested by other metrics validation researchers over the years and many are found to be weak [Meneely et al, 2010].
3. Her validation criteria are closely related to the representation condition, a property that many researchers cite as the most important category in metrics validation. Representation Condition criterion will however be covered as part of the work done by Kitchenham et al [Kitchenham et al, 1995].

So six of the seven criteria proposed by Weyuker were dropped as it appeared only in her paper. The seventh criteria i.e. Appropriate Granularity was included because it appeared in KITCH1995. Then the two lowest cited papers i.e. BRAIN1996 and FENTO2000 were studied to look for any opportunities to reduce the list.

From the paper BRAIN1996 i.e. "On the Application of Measurement Theory in Software Engineering", the validation criteria included:

1. A Priori Validity
2. External Validity
3. Theoretical Validity

From the paper FENTO2000 i.e. "Software Metrics: Roadmap", the validation criteria included:

1. Action-ability
2. Causal Model Validity

- **A Priori Validity** criterion was mentioned in both BRAIN1996 and FENTO1994 as shown in table 10. A metric has a priori validity if the attributes in association are specified in advance of finding a correlation. This is closely related to the Association criterion mentioned as part of the empirical validation criteria by Schneidewind i.e. SCHNE1992. So "A Priori Validity" criterion will be dropped for Association criterion.

- As said in section 2.3, **External validity** is also known as empirical validity. External/Empirical validation itself is a broad category of validation and it is related to some way with an external factor [El Emam, 2000]. Hence we drop this criterion from the list due to its abstractness. In addition, all of Schneidewind's six validation criteria in SCHNE1992 fall under empirical validation.

- As mentioned earlier, **theoretical/internal validation** is a broad category of validation and it is related to measurement theory axioms. Hence we drop this criterion from the list due to its abstractness.

- **Actionability** which appears in FENTO1004 allows for making empirically informed decisions. We will drop this for Schneidewind's empirical validation criteria list.

- **Causal Model Validity** is used to explain an external factor which belongs to External/Empirical validation. We will drop this for Schneidewind's empirical validation criteria list.

So these five criteria (from BRAIN1996 and FENTO2000) were also dropped along with the six dropped earlier from Weyuker. This brings the list of criteria to 17 (17 = 28 - 6 – 5) from three papers. The list of 17 validation criteria is as shown in the table 11 below.

Table 11: List of 17 Validation Criteria

SL #	Criteria	SCHNE1992	FENTO1994	KITCH1995
1	Appropriate Continuity			Y
2	Appropriate Granularity			Y
3	Association	Y	Y	
4	Attrubute Validity			Y
5	Discriminative Power	Y		
6	Dimensional Consistency			Y
7	Instrument Validity			Y
8	Predictability	Y	Y	
9	Process/Product Relevance	Y		
10	Protocol Validity			Y
11	Rank Consistency	Y		
12	Repeatability	Y		
13	Representation Condition		Y	Y
14	Scale Validity		Y	Y
15	Trackability	Y		
16	Underlying Theory Validity			Y
17	Unit Validity		Y	Y

2.6 Refining the Criteria List – Third Iteration

Each of the 17 criteria was again studied to see if any of the criteria can be dropped.

- Criterion # 9 in table 11 i.e. Process/Product Relevance from SCHNE1992 refers to the tailoring of the metric to specific products or processes. This criterion is not applicable as the goal of this research is not to derive metric/measure for specific products or processes.

- Criterion # 16 i.e. Underlying Theory Validity is associated with internal/theoretical and external/empirical validity [Kitchenham et al, 1995]. As external/empirical validity is a broad category of validation and it is related to some way with an external factor [El Emam, 2000], we drop this criterion from the list due to its abstractness.

This brings the total list of validity criteria to 15. These 15 criteria were categorized into seven theoretical and eight empirical validity measures given that researchers typically perform internal validation theoretically and external validation empirically. Coincidentally all seven criteria for the theoretical validation came from Kitchenham's paper [Kitchenham et al, 1995]. The final list of 15 criteria is as shown in table 12 below.

Table 12: List of 15 Validation Criteria

SL #	Criteria	Type	SCHNE1992	FENTO1994	KITCH1995
1	Appropriate Continuity	Theoritical			Y
2	Appropriate Granularity	Theoritical			Y
3	Association	Empirical	Y	Y	
4	Attribute Validity	Empirical			Y
5	Discriminative Power	Empirical	Y		
6	Dimensional Consistency	Theoritical			Y
7	Instrument Validity	Empirical			Y
8	Predictability	Empirical	Y	Y	
9	Protocol Validity	Theoritical			Y
10	Rank Consistency	Empirical	Y		
11	Repeatability	Empirical	Y		
12	Representation Condition	Theoritical		Y	Y
13	Scale Validity	Theoritical		Y	Y
14	Trackability	Empirical	Y		
15	Unit Validity	Theoritical		Y	Y

2.7 Addressing the Threats to Research Validity

Typical threats to the validity of research that is applicable for the proposed measurement framework includes:

- **Inappropriate selection of the measurement framework.** Given that there are numerous frameworks available, the measurement framework adopted should be appropriate for a software project.

- **Inappropriate selection of software project measures.** The measures selected should address the key questions of the project stakeholders for accurate project visibility.

- **Insufficient data collected to make valid conclusions.** This is dependent on randomly selecting the subjects from a good sample size where each individual member of the population has a known, non-zero chance of being selected as part of the sample. Random selection also avoids bias and homogeneity. Also if the sample size is not "large enough", the results are likely to be inaccurate.

- **Measurement framework applicable in too few contexts.** This can be addressed with the good sample size of diverse backgrounds and experiments under both controlled and uncontrolled settings.

- **Variation in data.** This could be the result of improper selection of target subjects. The data collected can be checked for reliability using the "Index of Variation" measure.

- **Rigor in the Experiment.** Experiments need replication and cross-validation at different times and conditions before the results can be interpreted with confidence.

2.8 Conclusion

Measurement frameworks can play a pivotal role in software projects if they are derived using the right research design principles ensuring reliability and validity. Reliability ensures consistent measurement each time the measurement framework is applied and this can be ensured by Test-Retest and Internal consistency test (using Cronbach's alpha). Validation is important because the acceptance of the measures in the measurement framework depends on whether the measures can be used as a predictor for the particular attribute. In the absence of standard validation criteria, 15 validation criteria have been identified from literature studies so that the proposed measurement framework can be theoretically and empirically validated.

Chapter 3: Construction of the Measurement Framework

3.1 Introduction

A software project is both technical and functional endeavor and a measurement framework should be a reflection of these aspects. In this regard, there is a consensus between metrics researchers and practitioners that three steps are needed to define and validate software metrics [Braind, 1995; Fenton, 2006; Soni, Shrivastava and Kumar, 2009]. These three steps are:

1. **Defining new measurement framework.**
   A measurement framework includes the measures and the implementation infrastructure needed for taking corrective actions.

2. **Validating them theoretically.**
   The main aim of theoretical validation is to establish whether the new measures are structurally sound and ensure they do not violate principles of measurement theory as measurement theory gives a clear definition of a measure by linking the real world and the world of numbers. [Braind, 1995; Pfleeger et al, 1997; Soni et al, 2009; Muketha et al, 2010].

3. **Validating them empirically.**
   The main aim of empirical validation is to establish whether the new measures are measuring what they are intended to measure and is complementary to the theoretical validation. Typically empirical validation of measures includes experiments, surveys and case studies [Braind, 1995; Soni et al, 2009].

The relationship between these three steps is as shown on figure 8 below.

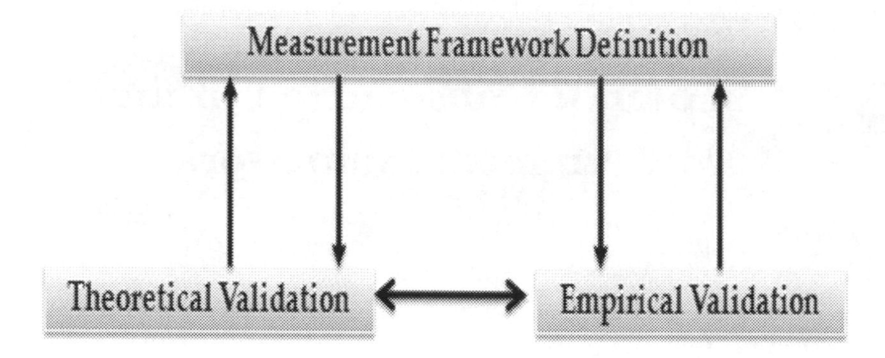

Figure 8: Metrics Validation

3.2 Basic Measurement Terminology

Metrology is one of the driving forces for the development and maturity of engineering disciplines including software engineering [Wang, 2003]. Though measurement is recording of past events it generally involves some predictive quantification of future effects with statistical interference. Essentially a measurement framework is composed of three building blocks – **Measure, Metric and Measurement**.

1. **Measure** can be defined as "to ascertain or appraise by comparing to a standard". It is the number or symbol assigned to an entity to characterize an attribute [Fenton and Pfleegar, 1997].

2. **A metric a.k.a key performance indicator (KPI) is** a quantitative measure of the degree to which the system possesses a given attribute. The Institute of Electrical and Electronics Engineers (IEEE) standard glossary of software engineering terminology defines metric as "A quantitative measure of the degree to which a system, component, or process possesses a given variable" [IEEE, 1990, pp 47].

The terms metric and measure have some overlap. Measure is used for more concrete or objective attributes while metric is for more abstract and subjective attributes. For example

Lines of code (LOC) is a measure. It is concrete and objective. Robustness or effectiveness is an important attribute that is abstract and hard to define objectively. This would be a metric. Similarly "body temperature" is a measure while "health" is a metric. In a line, measures are the base for metrics.

3. **Measurement is** the act or process of measuring. It is the process by which numbers or symbols are assigned to attributes of entities in the real world to characterize them according to clearly defined rules [Fenton and Pfleeger, 1997]. The measurement system has three elements.

    1. **Entity**.

        To measure, we must first determine the **entity** that needs to be measured**.** Entity is the level at which measurement is required. In the context of this research thesis, the measurement entity or object is the software project with three key elements or sub-entities, namely product, process and resources.

        i. The product is the final outcome of the project and has attributes such as functionality, usability, reliability, performance and security commonly known as "FURPS".
        ii. The process is how the product is developed. It has attributes such as size, productivity, effort, schedule, cost, quality, industry standards, government regulations, complexity, information flow, documentation to name a few.
        iii. Resources include people, hardware, software, office space and communication infrastructure that work on processes to produce the resulting product. For example, typical people attributes in a project are knowledge, skills and attitude popularly known as "KSA".

2. **Attributes.**

Attributes describe the characteristics of the entity that can be directly or indirectly measurable. Typical software project attributes include cost, complexity, size, schedule, quality etc. Attributes may be grouped into two main categories: internal and external as shown in table 13 [Kitchenham et al, 1995].

i. Internal attributes are measured directly from the entity and they are the easiest to measure. They are meta-attributes that may not be further divided into smaller elements. A good example of internal attribute is the size in a software program.

ii. On the other hand, external attributes are indirect and are measured as derivatives of internal attributes. An example of an external product attribute is quality, which may be measured by for example counting the number of defects in a development object.

Table 13: Attributes in a Software Project

Entities	**Attributes**	
	Internal	**External**
Products		
Specifications	Size, reuse, modularity, redundancy, functionality, syntactic correctness etc.	Comprehensibility, maintainability etc.
Design	Size, reuse, modularity, coupling, cohesiveness, inheritance, functionality etc.	Quality, complexity, maintainability etc.
Code	Size, reuse, modularity, coupling, functionality , algorithmic complexity, control-flow structures etc.	Reliability, usability, maintainability, reusability etc.

Test Data	Size, coverage level	Quality, reusability etc.
Processes		
Constructing specification	Time, Effort, Number of Specification changes	Quality, Cost, Stability
Detailed Design	Time, Effort, number of specification faults found etc.	Cost, Cost-effectiveness
Testing	Time, Effort, number of coding faults found etc.	Cost, Cost-effectiveness, stability
Resources		
Personnel	Age, billing rate, skill level etc.	Productivity, Experience, Intelligence etc.
Teams	Size, communication level, structuredness etc.	Productivity, Quality etc.
Organizations	Size, ISO certification, CMM level etc.	Maturity, Profitability etc.
Software	Price, Size etc.	Usability, Reliability etc.
Hardware	Price, Speed, Memory size etc.	Reliability, Performance etc
Offices	Floor area, Temperature, Lighting etc.	Comfort, Quality, etc.

3. **Mapping System or Rules**.

Rules (and scales) help in assigning values to the attributes where different rules lead to different scales. It is meaningless to say that a person's weight is 65 unless we know that we are talking about kilograms. In software project metrics for the number of LOC, the mapping system can be the physical lines of code.

3.3 Software Metrics and Measurement Frameworks

The history of software metrics is almost as old as the history of software engineering. The groundwork for software measurement was established mainly in the sixties and seventies, and further results have emerged since. Table 14 below shows the chronology of different measures proposed.

Table 14: Chronology of Software Metrics

Sl #	Metric	Dated to	Measurement Purpose
1	KLOC	1960s	Program Size
2	Defects	1960s	Quality
3	KLOC	1971	Program Complexity
4	McCabe	1976	Program Complexity
5	Halstead Metrics	1977	Program Complexity
6	Function Points	1979	Program Size
7	EVM Metrics	1987	Schedule and Cost
8	Mark II Function Points	1991	Program Size

(Though EVM has been around in the 1960s, it was included in the first PMBOK Guide in 1987)

The search team "Software Metrics" in Google gives over 170,000 results while the same search in Google Scholar gives more than 21,000 scholarly pieces such as thesis, books, and research papers. Amazon provides a list of over 100 books on software metrics. Most computing or software engineering courses world-wide now include some compulsory material on software metrics. In other words, judging by the number of books, journal articles, academic research projects, university courses and dedicated conferences, this subject area has now been accepted as part of the mainstream of software engineering.

Several measurement methodologies have emerged in recent years including some organizational frameworks that have an integrated measurement methodology embedded within them. Figure 9 below illustrates the chronology of measurement frameworks [https://goldpractice.thedacs.org/practices/gqm/index.php].

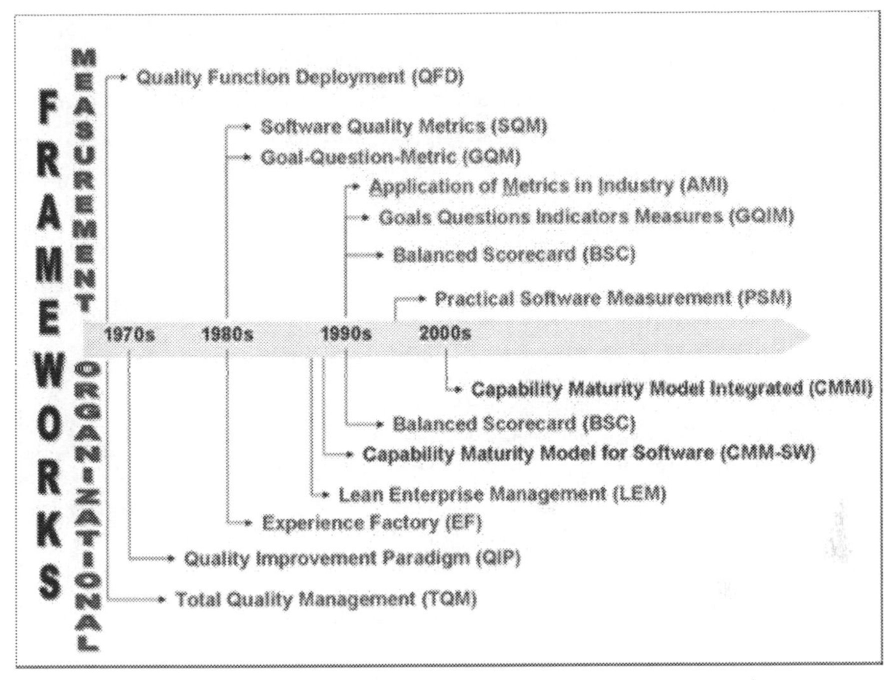

Figure 9: Chronology of Measurement Frameworks

Though several approaches to software measurement have been developed as shown in figure 9, the goal–oriented approaches for software measurement are commonly used because they use goals, objectives, strategies and other mechanisms to guide the choice of data to collect and analyze in a systematic way. According to Hall and Fenton, a metrics program established without clear and specific goals and objectives is almost certainly doomed to fail [Hall and Fenton, 1997]. According to Basili et al, most of the approaches attempting to align measurement with the business and software are combinations of three well-known approaches to measurement i.e., BSC, GQM, and PSM [Basli, Victor, Lindvall, Mikael, Regardie, Myrna, Seaman, Carolyn, Heidrich, Jens, Munch, Jurgen, Rombach, Deiter, Trendowicz, Adam, 2007].

In the coming sections we look at these three goal oriented measurement frameworks i.e., BSC, GQM, and PSM in detail and select the most suitable one for software projects.

3.3.1 Practical Software Measurement (PSM)

PSM is information driven measurement process which offers detailed guidance on software measurement to link issues, measurement categories, and measures [Florac, William, Park, Robert and Carleton, Anita, 1997]. It is essentially a issue-based measurement method, guiding software project managers to select, collect, define, analyze and report scientific software issues such as risks, defects etc.

The three concepts that form the foundation of PSM are [Card D.N and Jones C.L, 2003]:

1. **Information needs of Project Managers**.

   This drives the selection of the measures to influence the outcome of the project. This is derived from the goals the project manager seeks to achieve and the obstacles that hinder the achievement of these goals.

2. **The Measurement Information Model**.

   Once the goals and obstacles are identified, the next step is to define the Measurement Information Model. The Measurement Information Model a.k.a measurement construct provides a formal relationship between the information needs and the objective data to be collected; commonly called measures. The Measurement Information Model is as shown in the figure 10 [Card and Jones, 2003].

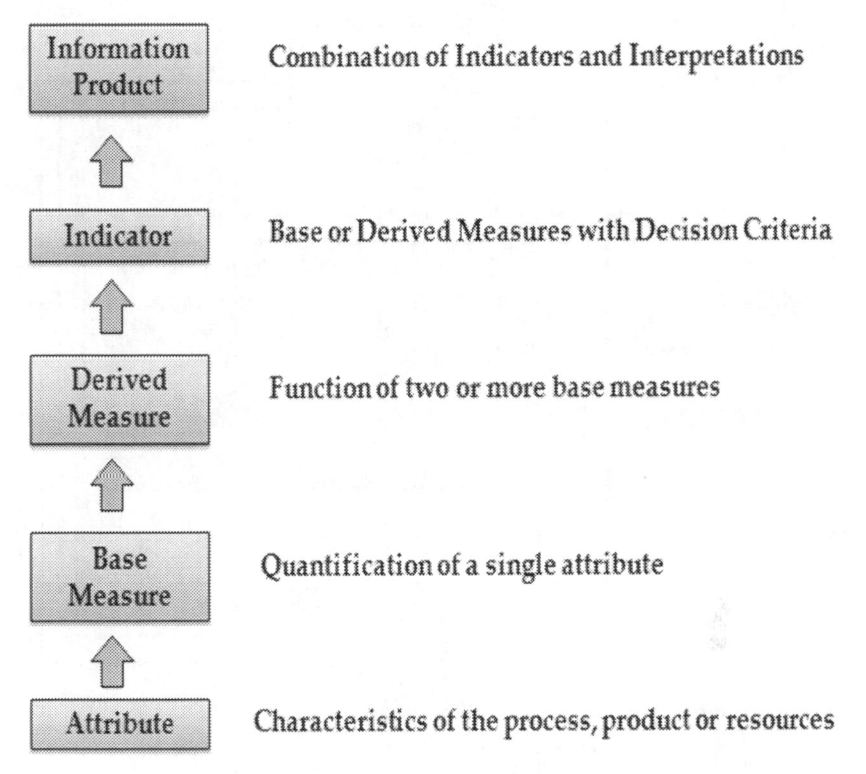

Figure 10: PSM Measurement Information Model

3. The Measurement Process Model.

An effective measurement process must address the selection of appropriate measures with an effective analysis of the corresponding data that is collected. The measurement process model tackles this through four iterative measurement activities: Establish, Plan, Perform, and Evaluate as shown in figure 11 below [Card and Jones, 2003].

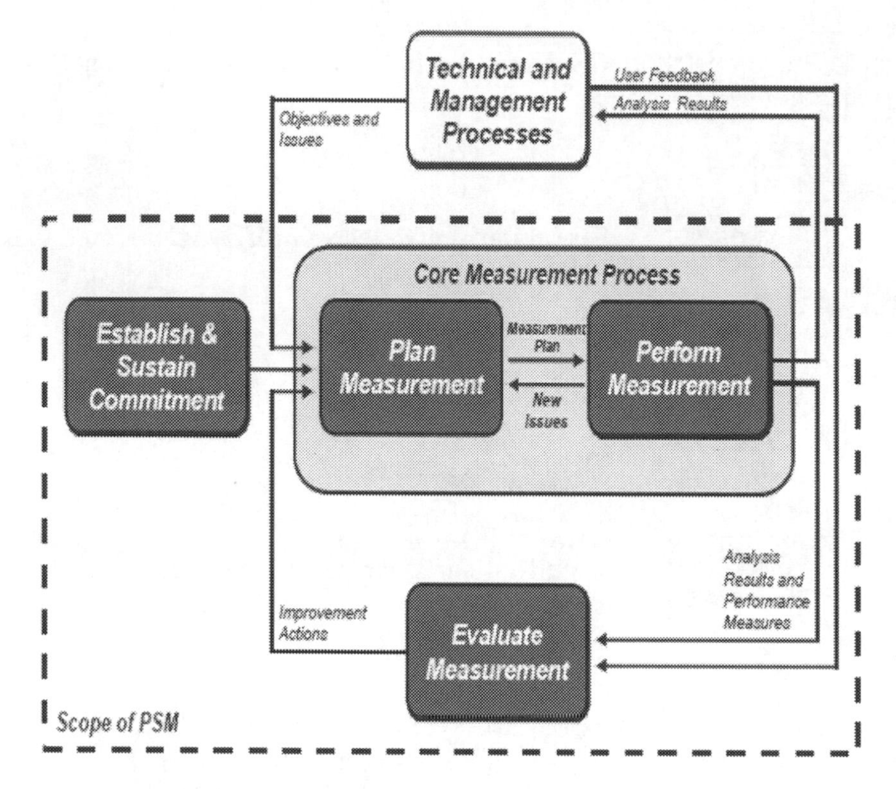

Figure 11: PSM Measurement Process Model

The key benefits of PSM are:

1. Project initiates the measurement activities.
2. Project characteristics guide the measure/metric selection.
3. Assesses measurement activities as part of the framework.

The short comings of PSM are:

1. PSM does not create the link to business goals. It sees measurement as project level activity only.
2. PSM is primarily a tool for project managers and not a measurement framework for all stakeholders in the project.

3. PSM focuses on using existing 75 measures/indicators repositories based on various information categories to know the project status. The challenges here are:

    - Implementing this complete metrics suite will consume a lot of time in data collection, analysis and implementation.
    - Moreover even if a small set of measures is carved out, it is still not appropriate for other stakeholders as the 75 measures are essentially designed for the project managers.

4. PSM was developed by US Department of Defense (DoD) based on experiences of software projects run by the Defense department. It is not a measurement framework coming out from the software industry practitioners.

3.3.2 Balanced Scorecard (BSC)

The Balanced scorecard (BSC) provides measures for project or organizational performance across four perspectives:

1. Financial
2. Customer
3. Internal business processes and
4. Learning and growth.

The term "scorecard" signifies quantified performance measures and "balanced" signifies that the system is balanced between:

- Short and long term objectives
- Financial and Non-financial measures
- Lagging and leading indicators
- Internal and external performance perspectives.

The characteristic of the BSC is the presentation of holistic measures where each measure is compared against a 'target' value within a single concise report. BSC does not provide a static set of measures, but serves as a framework for choosing measures, processes, and initiatives that are aligned with organizational vision, strategy and business goals. The idea is to create specific and actionable linkages among the four perspectives so that all project activities contribute to a unified vision and strategy to help projects/organizations align business activities, improve internal and external communications, and monitor organization performance [Kaplan and Norton, 1992]. In a well designed scorecard, the four perspectives form a chain of cause-effect relationships. For example learning and growth leads to better business/project processes resulting in higher user/customer satisfaction and thus a higher return on investment (ROI).

Within each of the four perspectives, BSC emphasis the following four parameters to be defined.

- **Objectives**: It is the goal to be achieved in that particular perspective
- **Measures**: How the progress for that particular objective will be measured
- **Targets**: The target is the goal sought for each measure
- **Initiatives**: The actions that will be taken to reach the target.

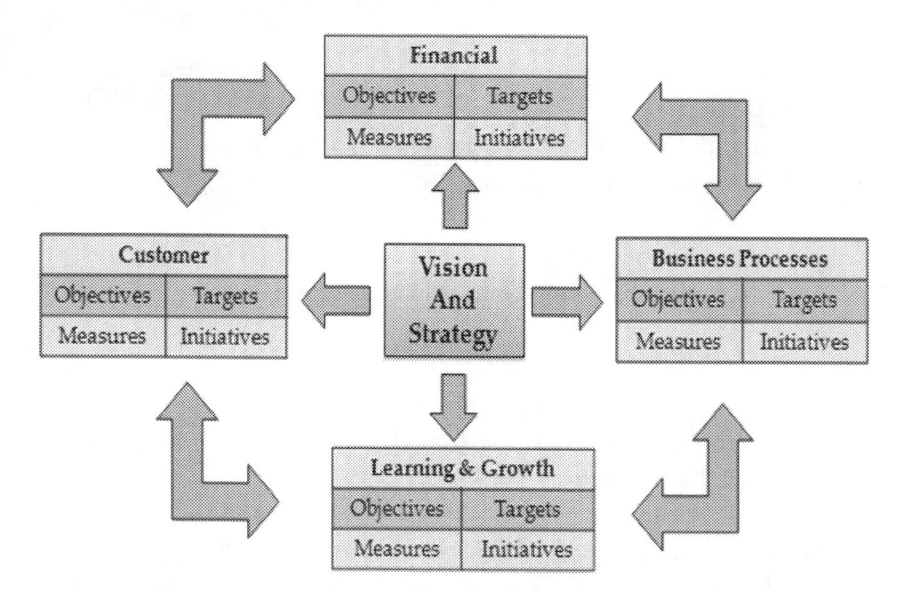

Figure 12: Balanced Scorecard

The below figure 12 shows the how the four perspectives and the four parameters fit together and interact.

The key strengths of BSC are:

1. Strong top management focus.

2. Links various company aspects under one management system. It aligns individual goals to the objectives of the organization/project.

3. Links measurement to the organization's vision and goals.

The disadvantages of BSC are:

1. BSC is focused on upper management needs and essentially involves management stakeholders.

2. The four areas in BSC are abstract and do not mention any specific measures. It gives very little support for project level measurement definition.

3. BSC is generic and not specific to software projects.

3.3.3 Goal-Question-Metric (GQM)

Goal-Question-Metric (GQM) method was formulated by Victor Basili, Caldiera Gianluigi and Deiter Rombach in 1994 [Basili, Gianluigi and Rombach, 1994]. The GQM approach provides a method for defining goals, refining them into questions and formulating metrics for data collection, analysis and decision making. GQM focuses on getting the right people involved at all levels to ensure the right goals and metrics are identified to get buy in from all the stakeholders. GQM also seeks to involve the stakeholder in the entire process and focuses on the human and cultural issues for a successful measurement program.

The strengths of GQM are:

1. Makes a visible link from measurement goals to the data collected.

2. Creates a detailed measurement plan.

3. Gives a model for analyzing collected measurement.
4. Involves all stakeholders in measurement definition, implementation and analysis.

The weaknesses of GQM are:

1. GQM doesn't provide explicit support for integrating its software measurement model with the larger element of the organization such as business goals, strategies and assumptions.
2. Creates too much flexibility and not enough guidance. For instances in formulating the questions, GQM does not provide guidance on when to terminate.
3. GQM does not bring uniqueness. For the same goal(s) and questions of the stakeholders there could be more that one set of measures.

3.4 Which Measurement Framework to Adopt?

Each of the three goal-oriented measurement frameworks (GMF) has their own pros-and-cons. The challenge is which measurement framework to adopt for software projects. To summarize, BSC focuses on the business vision from four concrete perspectives, GQM is useful to define the strategic goals reflecting stakeholder needs and PSM provides a set of pre-defined measures. The below table 15 illustrates the key differences between the three goal-oriented approaches [Basili, 2003; Sirvio, 2003].

Table 15: Comparison of Goal Oriented Measurement Frameworks

Sl #	Critical Success Factor Questions	GQM	PSM	BSC
1	Does the method support participation of all affected parties?	X	X	X
2	Does the method support co-operation with software engineers?	X	X	
3	Does the method support planning and carrying out training as part of the initiative?		X	
4	Does the method support commitment of top managers?			X
5	Does the method support commitment of middle managers?			X
6	Does the method support commitment of software engineers?			
7	Does the method support that improved solutions are developed on a case-by-case basis?	X	X	X
8	Does the method support that the current status of processes is clarified?	X		
9	Does the method support that the link between business and improvement goals is established?			X
10	Does the method ensure measurement goals are based on needs and well understood?	X		X
11	Does the method ensure that detailed measurement plan is generated?	X	X	X
12	Does the method support developed solutions are tested in a pilot project?	X	X	
13	Does the method ensure that practical support is always available for development projects?	X		
14	Does the method support using metrics in monitoring improvement actions and results?	X	X	X
15	Does the method support sustainability of an improvement initiative?	X	X	X

From table 15, GQM is more suited for software projects. While GQM approach to measurement remains popular with researchers, PSM is becoming the widely practices approach to management by fact by software project managers [Card and Jones, 2003]. Moreover Google Scholar gives about 4000 hits for GQM compared to approximately 600 hits for PSM (as on May 2011) showing the popularity of GQM over PSM in the research community.

The three goal oriented measurement frameworks (GMF) were further validated with a poll in LinkedIn and Twitter with software industry practitioners across the globe. The responses from 82 industry practitioners are as shown in the figure 13. The "Others" category (23 responses; 28%) is to segregate the responses from people who have a clear preferences of the three GMFs to those who do not have preference on the three goal-measurement frameworks. The "Other" responses include those who do not have an opinion, prefer other measurement frameworks such as Planguage, believe in a combination of the three frameworks etc.

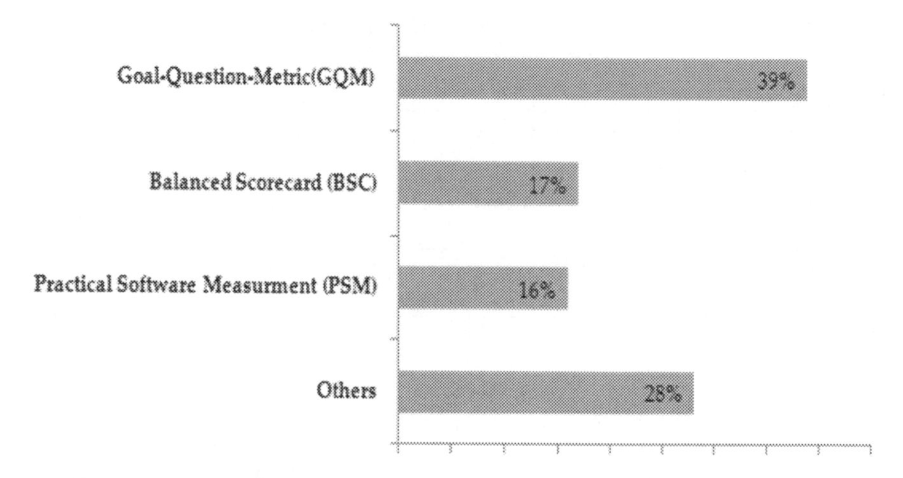

Figure 13: Preferences of GMFs from Industry Practitioners

However as the objective is to find the best goal oriented measurement framework from the three models, the responses of 59 participants (after excluding those who voted for the "Others" category) was considered and it is as shown in figure 14. This clearly indicates that GQM is the preferred model (with 32 of the 59 respondents choosing GQM) of the three goal measurement frameworks amongst the industry practitioners.

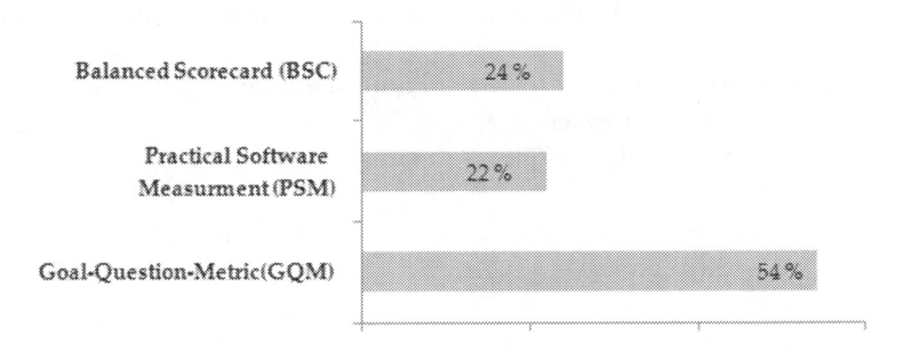

Figure 14: Preferences of Three GMFs from Industry Practitioners

The research and poll show that GQM is the most suitable model for software measurements as aligns with the organizational goals given that goals shape the targets for measurement, questions support the accomplishment of the goals and metrics provide answers to the questions. According to Berander and Jönsson, GQM has become a de-facto standard for the definition of measurement frameworks [Berander and Jönsson, 2006; Fenton, 2006]. Hence the GQM model will be adopted for deriving the software project measures in this research thesis.

3.5 The GQM Approach

The GQM approach is based upon the assumption that for an organization to measure in a purposeful way it must first specify the goals for itself and its projects, then it must trace those goals to the data that are intended to define those goals operationally, and finally provide a framework for interpreting the data with respect to the stated goals. The GQM model is characterized by two processes:

1. Top-down refinement of measurement goals into questions and then into metrics
2. Bottom-up analysis and interpretation of the collected data.

GQM is based on eight principles namely:

1. **Goal-driven:** Define measurement goals in line with the project goals.
2. **Context-sensitive:** Consider context/environment when defining measurement goals.
3. **Top-down:** Refine goals top-down into measures via questions.
4. **Documented:** Document measurement goals and their refinement explicitly.
5. **Bottom-up:** Analyze and interpret the collected data bottom-up in the context of the goal.
6. **People-oriented:** Actively involve all stakeholders in the measurement program.
7. **Sustained:** Measure for systematic and continuous software process improvement (SPI).
8. **Reuse-oriented:** Describe the context to facilitate packaging and reuse of knowledge gained.

The result of the application of the GQM approach is a measurement system targeting a particular set of issues and a set of rules for the interpretation of the measurement data. The resulting measurement model has three levels:

1. **Conceptual level (Goal)**

   A goal describes the purpose of the measurement and is defined for an object with respect to various models of quality, from various points of view and relative to a particular environment. The development of a goal is based on three basic sources of information [Basili et al, 1994].

   i. Strategy. This includes the organization policies and business plans.
   ii. Scope. This is the description of the measurement we want to perform.
   iii. Relevance. Given that, not all issues and processes are relevant for all viewpoints in an organization, a relevancy analysis must be carried out before completing the list

of goals to make sure that the goals defined have the necessary relevancy.

2. **Operational level (Question)**

   A set of questions is used to define models of the object of study and then focus on that object to characterize the assessment or achievement of a specific goal. The questions generated should define the goals in a quantifiable way. The questions are at the operational level and help to clarify and refine the goal and to capture the variation of understanding of the goals that exists among the different stakeholders.

3. **Quantitative level (Metric)**

   A set of metrics is associated with every question in order to answer it in a measurable or quantitative way. The measures specified should be collected to answer the questions and track process and product conformance to the goals.

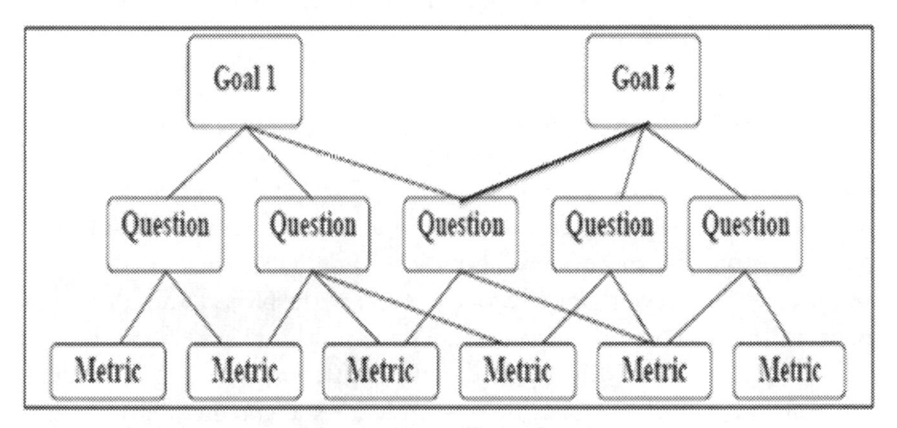

Figure 15: GQM Model

As illustrated in the figure 15 above [Basili et al, 1994, pp 3], the mapping among goals, questions and metrics is not one-to-one. For each goal, there can be several questions and the same question can be linked to multiple goals as appropriate. For each question, there can be multiple metrics, and some metrics may be applicable to more than one question. Adherence to and preservation of this hierarchical structure helps ensure that the measurement program focuses on the right metrics and that we avoid extra work associated with collecting metrics that are not really needed.

3.6 Building GQM based Measurement Framework.

In section 3.4, we identified GQM as the most suitable measurement framework for software projects. In this section we build a stakeholder driven measurement framework in six steps applying the concepts of GQM.

Step 1 is categorizing the stakeholders into three groups and identifying the stakeholder criteria for project success. In step 2, we articulate a generic goal statement from the stakeholder perspective. We translate the goal to quantifiable questions in step 3. Given that step 3 can potentially give rise to a huge list of questions, step 4 prioritizes those questions using PCA technique. Once we have the final list of questions, step 5 maps the questions to the respective project attributes. Finally, step 6 derives measures using those attributes.

3.6.1 Step 1 - Conduct Stakeholder Analysis

Software project management is generally stakeholder-driven. Hence we need to identify the stakeholders and their interests or stakes— that determine the success of the project and reflect their goals in the measurement framework for effectively tracking the project and ultimately ensuring the success of the project. However the views of the project stakeholders are often diverse and reflect different political, technical, functional and organizational interests. Hence stakeholder analysis helps to align the stakeholder needs to the business and project goals. It is not important whether the business goals are developed under the umbrella of GQM, or as a function of strategic planning. But business goals must exist and the project goals should be tied to the business goals for clarity in stakeholder needs.

Stakeholders can be of any form, size and capacity. They can be individuals, organizations, or unorganized groups. However stakeholder identification and analysis must preferably be done throughout the project and the stakeholder needs reflecting the overall project success must be captured in through the measures [Rad and Ginger, 2005]. Four major attributes in stakeholder analysis are:

1. The stakeholders' position in the project such as Sponsor, Project Manager, Developer, User etc.
2. The level of influence (power) they hold.
3. The level of interest they have in the project.
4. The group/coalition to which they are associated with.

The intensity of these attributes amongst the stakeholders varies according to the project situations. Broadly, these attributes signal the capability the stakeholder has to block, hamper or promote the project, join with others to form a coalition of support or opposition, and lead the direction/discussion of the project. Stakeholder analysis therefore provides a detailed understanding of the political, economic, and social impact of the project on the affected interested groups. In this scenario, based on the work of Turner et al [Turner et al, 2009], below are the three generic categories of project stakeholders and their primary concerns for project success:

1. **Initiators**. This group includes stakeholders such as project sponsor, management team etc who bring the project into existence. All projects have initiators– these are the business champions who see a need for change, provide resources, and have the authority to make something happen. Without them, the project would not have been proposed. Their primary concerns would be on return on investment (ROI), total cost of ownership (TCO) and adherence to schedule and cost. ROI and TCO would be a function of quality of the software product.

2. **Implementers**. This group includes stakeholders such as the project manager, business analysts, developers and testers including the suppliers who build the project. They translate the project visions and plans into reality. Their primary concerns would be on size/features and adherence to schedule and cost.

3. **Beneficiaries**. This group is the actual users of the software applications who desire a functional, reliable, user-friendly and a secure software application. Their primary concerns would be on size/features and quality.

Table 16 below summarizes the project stakeholders and their primary concerns for project success.

Table 16: Primary concerns of project

Project Stakeholders	**Primary concerns**
Initiators	Schedule, cost and quality
Implementers	Size/features ,schedule and cost
Beneficiaries	Size/features and quality

Categorizing the stakeholders into three groups also meets the fundamental step in measurement – classification. A good classification procedure not only produces accurate classifiers (within the limits of the data) but that is also *provides insight and understanding into the predictive structure of the data* [Breiman L, Friedman J.H, Olshen R.A, Stone C.J, 1984, pp7]. A proper analysis and classification of the stakeholders will help to understand the right players, their importance in the project, their needs and communication strategies. According to Putnam and Myers, not only must we establish measures, their collection and accessibility, but we must also communicate the nature of metrics based management to these stakeholders [Putnam and Myers, 2002].

3.6.2 Step 2 – Formulate the Stakeholders Goal(s)

While the stakeholders' needs are normally stated as business goals, the goals from the GQM perspective are the measurement goals. To translate the business goals to measurement goals, *the business goals must first be decomposed and refined to a point where meaningful entities, purposes, perspectives, and environments can be identified. To aid this exercise, the GQM model provides a* template with five dimensions namely: object, purpose, focus, viewpoint and environment [Basili et al, 1994].

1. **Object** is an entity that will be measured and analyzed. Objects of measurement in a software project are:
   - **Products**
     A software product is typically a suite of functionalities built to be used by many customers, businesses or consumers.This includes deliverables that are produced during the system lifecycle. Examples are specifications, designs, programs, standards, test scripts, training aids etc.

   - **Processes**
     An essential aspect of software engineering is the discipline required for a group of people to work together cooperatively to solve a common problem. Humphrey defines the term *software process* as "a sequence of steps required to develop or maintain software" [Humphrey, 1995, pp 4, 441-459]. In the context of the software development lifecycle (SDLC), processes include software related activities normally associated with time. Examples include specifying, designing, coding, testing, managing, controlling, deploying etc.

   - **Resources**
     Resources include items used by processes in order to produce the output which is the software product. Examples include personnel, hardware, software, tools, office space etc.

2. The **purpose** expresses what will be measured in the object. It is the motivation behind the goal i.e. the "why" the goal exists. It is the reason to achieve the goal. For example purpose can be better control, improvement etc.

3. The **focus** is the particular attribute of the object that will be analyzed in the measurement framework.

4. The **viewpoint** provides information about the people who will interpret and use the metrics.

5. Finally, the **environment** is the context in which the measurement study will be performed.

There can however be multiple goals for one measurement entity. In the context of this research, the measurement goal statement encompassing the five dimensions of the GQM framework for a software project will be - Track the software project objectively (**object**) to deliver the scope successfully (**purpose**) with respect to size, complexity, schedule, cost and quality (**focus**) from the viewpoint of the stakeholders (**viewpoint**) given the high failure in software projects because of a lack of management visibility (**environment or context**).

Once the goal statement is defined, it can be further refined with compliance to the "SMART" template i.e.

- **Specific**:
  When goals are specific, they convey precisely what is expected, why is it important, who is involved and which attributes are hold the key for progress and success. In other words, the goal statement should describe specifically the desired result in the project.

- **Meaningful**:
  When goals are meaningful, they bring concrete criteria for measuring progress in determining whether the objective is accomplished or not.

- **Attainable**:
  Goals must be realistic and attainable. Goals that are set too high or too low often become meaningless. While a high goal set becomes difficult to achieve, a low goal results in poor motivation. So it is crucial to have the right goals so that it can also be used as a tool to address capacity.

- **Relevant**:
  The goal defined should be relevant to the issues in the project. For instance setting a sign-off process for the software requirements specification (SRS) document at 100% where

the requirements are volatile and project stakeholders are not aligned will not be a relevant goal.

- **Timely**:
  The goal statement should be grounded within the project time frame. In other words, it is not simply, "reduce project costs by 15 percent". It should be "reduce project cost by 15 percent in the next three months." This is the final anchor in making the goal statement real and tangible.

The refinement of the goal statement is as shown in table 17 below.

Table 17: Refining the Goal Statement

Initial Goal Statement (GQM Template)	**Refined Goal Statement (SMART Template)**
Track the software project objectively to deliver the scope successfully with respect to size, complexity, schedule, cost and quality from the viewpoint of the stakeholders given the high failure in software projects because of a lack of management visibility.	Track the software project objectively **during the duration** to deliver the scope successfully with respect to size, complexity, schedule, cost and quality from the viewpoint of the stakeholders given the high failure in software projects because of a lack of management visibility.

3.6.3 Step 3 - Translate the Goals to Quantifiable Questions

Moving from measurement goals to quantifiable questions is a crucial phase *in the GQM model*. The entity-question list is the recommended tool to help identify and frame quantifiable questions [Park et al, 1996]. To translate the business or measurement goals of the stakeholder to quantifiable questions, a two phase "Define-Refine" process was applied.

Phase 1: Define a baseline question list at the lowest level of the entity

For decades, journalists have used a proven approach called the 5WH (Who, What, When, Where, Why, and How) to answer the questions newspaper readers commonly want answered in the article. The underlying principle behind the 5WH questions is that each question elicits a factual answer, and none of them can have a simple "yes" or "no" answer. This strategy was leveraged to draw a list of base-line questions for each of the sub-entities (process, product and resources) of the entity (project) to help the stakeholders in better understanding and realizing their project goals. So the base line questions for the three sub entities of the entity are as shown in table 18 below.

Table 18: Base Line Questions

	Product	Process	Resources
Who	Who will use the product?	Who/What is the critical path ?	Who are the stakeholders?
What	What is the delivery status? What is the functionality to be delivered?	What is the scope? What factors are critical to quality (CTQ) ?	What resources do we need?
When	When is the product ready for deployment?	When is the project delivery due?	When will the resource consumption start and end?
Why	Why is the product complex?	Why/Where are the risks ?	Why do constraints, assumptions and dependencies exist?
Where	Where are the impact of changes ?	Where can the CTQs be optimized?	Where is the buffer allocated?
How	How stable are the requirements?	How much is the budget ? How is the current level of quality?	How is the "Waste" in the project determined?

Phase 2: Refine the list with questions at the next level above.

Stepping back, it was checked if anything was missed in the base-line question list in table 18 by asking questions in level "N+1" for the entity at level 'N'.

- What determines success of the entity at level N?
- How is the stability of the process?
- Where is the capability constrained?
- What factors can be controlled?
- Where do the early warnings signals come from?

To cover all the levels, these two phases i.e. define and refine were executed starting from the lowest level of measurement entity and moving up iteratively until the final entity (i.e. software project) was reached as shown in the figure 16 below.

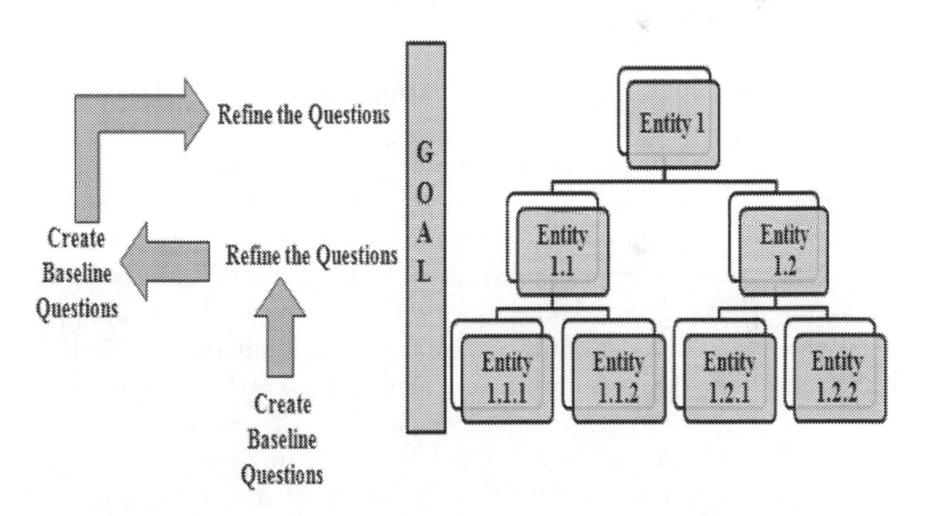

Figure 16: Formulating Questions in GQM

For example, program code (entity 1.1.1) could be the sub entity of the product (entity 1.1) and the product can be the sub entity of the project (entity 1) as shown in figure 17.

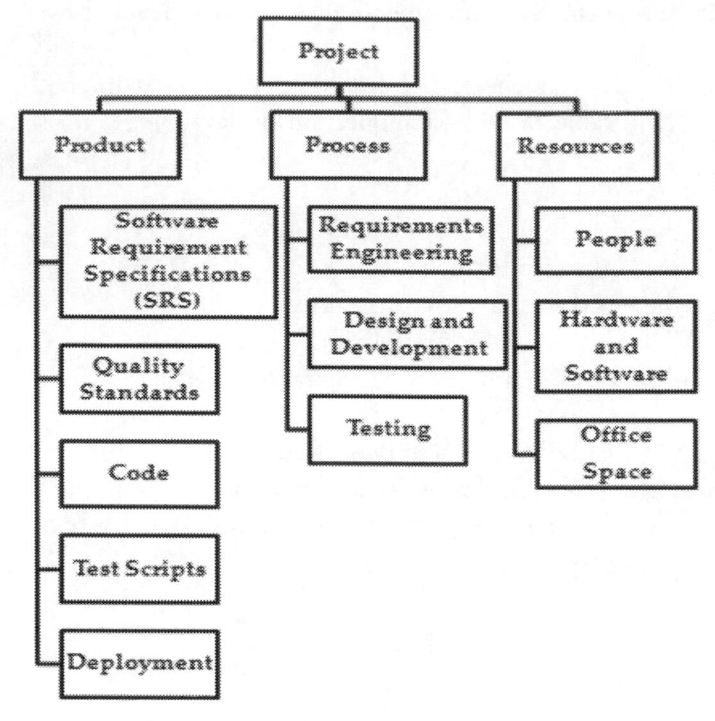

Figure 17: Example on Formulating Questions in GQM

Getting the right level of abstraction in the questions can be challenging. If questions are too abstract, the relationship between the measure and the question may be muddied. If they are too detailed, it becomes more difficult to get a clear interpretation of the goal. So for getting a complete list of questions with the right clarity, multiple iterations of the "Define-Refine" process were carried out and 14 questions stakeholders can have in a project were identified. The 14 questions are:

1. How to estimate the size of the project before development?
2. How to know the functional size or size of the project after development?
3. What is total estimated effort? Or how much will this project cost?
4. What is the complexity of the deliverables?
5. What is the estimated duration of this project?
6. What is the current stage of the project?

7. What is the productivity or what are the schedule and effort variances?
8. What is the delivery effectiveness in the project?
9. What is the critical path?
10. What is the current level of quality?
11. What is the impact and amount of re-work/cost of quality?
12. How is the requirements volatility?
13. What is the mean time to failure (MTTF) or what is the process stability?
14. What are the risk levels with respect to schedule, cost and quality?

3.6.4 Step 4 – Prioritize the Questions

The above 14 questions identified were prioritized by a sample set of stakeholders using the Paired Comparison Analysis (PCA) technique to work out the importance of every question relative to each other. This technique helps to set priorities to solve important problems in the absence of historical data. The objective of PCA is twin-fold.

1. To prioritize the 14 questions.
2. Explore if any of these 14 questions can be eliminated.

Fifteen stakeholders – three initiators, seven implementers and five beneficiaries were selected from different companies and from different countries and were asked to give the importance of the 14 questions relative to each other by giving a score from 0 (no difference) to 3 (major difference). From those 15 tables, the final value in each cell was selected based on the common responses. For example, in Q4 v/s Q7 comparison, eleven of the 15 stakeholders had marked Q7 as more important than Q4 and the average score difference between Q4 and Q7 after rounding to the nearest integer was 2 giving the final cell value as (7, 2). The first co-ordinate is the question preferred and the second co-ordinate is the preference score. This exercise was conducted for all the cells and the final values are shown in the table 19 below.

Table 19: Summarized Data from Paired Comparison Analysis (PCA)

(NA – Not Applicable; ND – No Difference; DP - Duplicate)

	Q1	Q2	Q3	Q4	Q5	Q6	Q7	Q8	Q9	Q10	Q11	Q12	Q13	Q14
Q1	NA	ND	1,1	1,2	1,1	6,2	1,1	8,2	1,3	1,2	1,1	1,3	1,2	14,2
Q2	DP	NA	2,1	ND	2,2	6,2	2,1	2,2	2,2	10,2	11,2	2,3	13,2	14,1
Q3	DP	DP	NA	ND	3,1	6,3	7,3	3,1	3,3	3,1	3,2	3,3	3,3	3,2
Q4	DP	DP	DP	NA	4,1	4,2	7,2	4,2	9,3	4,3	4,1	4,2	4,2	14,3
Q5	DP	DP	DP	DP	NA	5,2	5,2	5,1	5,3	10,2	11,2	5,3	13,2	5,2
Q6	DP	DP	DP	DP	DP	NA	6,1	6,2	6,3	10,2	6,2	6,1	13,3	14,2
Q7	DP	DP	DP	DP	DP	DP	NA	8,1	7,3	7,1	11,1	7,2	7,2	7,1
Q8	DP	DP	DP	DP	DP	DP	DP	NA	8,3	8,2	8,1	8,3	8,1	ND
Q9	DP	DP	DP	DP	DP	DP	DP	DP	NA	10,3	ND	9,1	ND	14,3
Q10	DP	DP	DP	DP	DP	DP	DP	DP	DP	NA	10,2	10,2	13,3	10,1
Q11	DP	DP	DP	DP	DP	DP	DP	DP	DP	DP	NA	11,2	11,2	11,3
Q12	DP	DP	DP	DP	DP	DP	DP	DP	DP	DP	DP	NA	12,2	14,2
Q13	DP	DP	DP	DP	DP	DP	DP	DP	DP	DP	DP	DP	NA	13,2
Q14	DP	DP	DP	DP	DP	DP	DP	DP	DP	DP	DP	DP	DP	NA

After adding up all the values for the 14 questions and converting each into a percentage of the total value of preferences i.e.166, we get the preference total for each of the 14 questions as shown below.

- Q1 = 16 (9.9%)
- Q2 = 14 (8.5%)
- Q3 = 16 (9.9%)
- Q4 = 13 (7.8%)
- Q5 = 13 (7.8%)
- Q6 = 16 (9.9%)
- Q7 = 13 (7.8%)
- Q8 = 12(7.3%)
- Q9 = 1 (0.6%)
- Q10 = 14 (8.4%)
- Q11 = 12 (7.3%)
- Q12 = 1 (0.6%)
- Q13 = 12 (7.3%)
- Q14 = 13 (7.8%)

This indicates that Q9 (What is the critical path?) and Q12 (How is the requirements volatility?) are not accepted by this set of stakeholders. Upon close analysis these two questions were "camouflaged" in the remaining 12 questions. For instance, Q9 on critical path (tasks that determine the end date in the project) was similar to Q5 on project duration. Q12 on requirements volatility in essence questions the impact of schedule, cost and quality in the project and is covered on questions pertaining to questions on risk levels and productivity. So 12 questions to know the project status that were finally selected were:

1. How to estimate the size of the project before development?
2. How to know the functional size or size of the project after development?
3. What is total estimated effort? Or how much will this project cost?
4. What is the complexity of the deliverables?
5. What is the estimated duration of this project?
6. What is the current stage of the project?
7. What is the productivity or what are the schedule and effort variances?
8. What is the delivery effectiveness in the project?
9. What is the current level of quality?
10. What is the impact and amount of re-work/cost of quality?
11. What is the mean time to failure (MTTF) or what is the process stability?
12. What are the risk levels with respect to schedule, cost and quality?

3.6.5 Step 5 - Derive Attributes from the Questions

The 12 questions have to be linked with attributes (characteristics of objects i.e. project). The attributes have to be selected because any measure π is a three-tuple $\prod = (\alpha, \Omega, \mu)$ where, α = Attribute to be measured, Ω = Measurement scale, and μ= Unit of measure [Wang, 2003]. Each of the 12 questions was associated with one of the following six project attributes where an attribute is defined as a feature or property of an entity [Fenton and Pfleeger, 1997]. The six attributes are:

- Size (physical and functional)
- Complexity or Maintenance

- Cost
- Schedule
- Stability (process and product)
- Quality Achieving Velocity

Stability (of the process and product) and the Quality Achieving Velocity can be combined into one attribute group - quality. This translates to five main project attributes – size, complexity, cost, schedule and quality.

3.6.6 Step 6 - Derive Measures from the Attributes

Each of the twelve questions (and their attributes) needs to be associated with the appropriate metric(s). In GQM context, the term metric is loosely defined; it can mean a base measure, a derived measure or a composite of measures. But the goal statement is to define measures in such a way that they provide objective information to answer to the 12 questions on the five attributes. Given that there are hundreds of measures available, five important factors were considered while associating the question (with the attribute) in the selection of appropriate measures [Basili et al, 1994; Wang, 2003].

1. **Application at all phases of the SDLC.**

   Measures selected could be applied in all the SDLC phases (i.e. requirement elicitation, development, testing and deployment phases). Focusing on the entire SDLC, rather than just on one phase, gives a comprehensive knowledge needed to enhance software quality. According to Tom DeMarco, "If you measure exactly the work of design, for instance, and then don't measure the coding at all, people will soon catch on. Their conscious or unconscious reaction will be to push as much of the work as possible into the unmeasured activity" [DeMarco, 1986].

2. **Objectivity.**

   Objective measures are preferred over subjective measures as they bring consistency in the measurement process. An objective measure is a measure where there is little scope for human judgment in the measurement value and is therefore primarily dependent on the object that is being measured. This brings consistency in the measurement process as an objective measure can be measured several times and the same value can be obtained within the measurement error. In this backdrop, given that a measure π is defined as three-tuple $\prod = (\alpha, \Omega, \mu)$ where,

α = One of the five attribute (size, complexity, schedule, cost and quality) to be measured.
Ω = Measurement scale
μ= Unit of measure

Ω and μ make the measure objective.

3. Level of Measurement/Scale types.

A measurement scale is a quantitative metrical yardstick that provides measuring unit and scope for a specific type of attribute of objects [Wang, 2008]. The scale type for a measure determines:

- Which statements about the measure are meaningful
- Which statistical operations can be applied to the data

There are four scale types namely - Nominal, Ordinal, Interval and Ratio; each with a distinct set of properties. At the nominal level, you have only categorization. At the ordinal level, one has the knowledge about the order of the categories along with categorization. With interval scales, you not only classify and order your measure but also define how much the categories differ one from another. A ratio scale has all of these three characteristics as well as a non-arbitrary, or true, zero value. The amount of information present is as shown in the table 20 below and is the highest for the ratio scale measures.

Table 20: Information Level in Measurement Scales

Level of Measurement	Categorization	Order + Categorization	Set Intervals + Order + Categorization	True Zero + Set Intervals + Order + Categorization
Ratio	X	X	X	X
Interval	X	X	X	
Ordinal	X	X		
Nominal	X			

The key message from this table is each higher level of measurement scale provides additional information. In addition, most statistical techniques make sense when used with interval or ration level measurement, because most of these statistical techniques involve taking differences among scores or sums of scores. Hence the measures which have the highest level of information were preferred over others. In addition, the scale type of the objective measures are either interval or ratio scale [Wohlin et al, 2000]

4. **Availability in existing tools.**

A measure that is already proven i.e. well researched, validated and implemented in tools such as Visual studio, Eclipse and Microsoft project 2003 is selected over others for:

- Easier and quicker implementation,
- Cost effectiveness and
- Lack of repetition.

5. **Flexibility in Implementation.**

The measures identified should be flexible enough given that refinement and adaptation would be needed during implementation. The three main flexibility factors (FF) are:

- FF1: Derivation/calculation of the measures in more than one way.
    1. FF1 is low (score is 1) if the measure has only one way of determination and

2. FF2 is high (score is 2) if the measure has only more than one way of getting determined.

- FF2: Measurement value available on demand at any stage in the project.
    1. FF2 is low (score is 1) if the measurement value takes more than 8 business hours to calculate (assumption is that the relevant data is available) for the lowest element in the WBS where the project is tracked.
    2. FF2 is high (score is 2) if the measurement value takes less than 8 business hours to calculate for the lowest element in the WBS where the project is tracked.

- FF3: Measure has a high repeatability with a low range of measurement error.
    1. FF3 is low (score is 1) if the repeatability is low and the range of measurement error is high.
    2. FF3 is high (score is 2) if the repeatability is high and the range of measurement error is low.

From the above five criteria described, the answers to the 12 questions (and five attributes) came from eight measures namely:

1. **Lines of Code (LOC).**
   It is the "physical" count of any programming statement without the blank or comment line [Wolverton, 1974].

2. **Function Points (FP).**
   It is the unit of measurement to express the amount of business functionality provided to the business user [Albrecht, 1979].
3. **McCabe's Cyclomatic Complexity (v (G))**.
   It measures the technical or system complexity by counting the available decision paths in the program [McCabe, 1976].

While LOC and FP reflect the size, v (G) is indicates the complexity of the product. The project schedule and cost attributes/measures come from earned value management (EVM). EVM measures the project progress in an objective manner combining measurements of scope, schedule, and cost in a single integrated system [Fleming and Koppelman, 2000]. The two measures from EVM are:

1. **Schedule Performance Index (SPI).**
   It is an index showing the efficiency of the time utilized in the project. SPI indicates how much ahead or behind schedule the project is [Fleming and Koppelman, 2000].
2. **Cost Performance Index (CPI).**
   It shows the efficiency of the utilization of the resources/ budget in the project. CPI indicates how much over or under budget the project is [Fleming and Koppelman, 2000].

Quality in the project can be achieved and improved by identifying and resolving the defects. As software quality is a multi-dimensional notion, three measures on quality are selected to address the relevant questions.

1. **Sigma Level (Cpk).**
   This indicates the effectiveness or stability of the entire software project delivery process [Pyzdek, 2000]. A higher Cpk indicates a process that is less prone to defects enabling us to achieve repeatable results.
2. **Defect Density (DD).**
   It compares the number of defects in various software components reflecting the stability of different components.
3. **Defect Removal Efficiency (DRE).**
   It indicates the velocity at which quality is achieved i.e. the rate at which defects are resolved.

The application of the five criteria in deriving the eight measures is as shown in table 21.

Table 21: Application of the five criteria on the eight measures

	SDLC Phases				Measure	Scale	Availability in	Flexibility			
Measure\Criteria	Req	Design	Dev	Testing	Type	Type	Commercial Tools	FF1	FF2	FF3	FFavg
LOC	NA	X	X	X	Objective	Ratio	Development Editors	2	2	2	2.0
FP	X	X	X	X	Objective	Ratio	Development Editors	2	2	1	1.7
v(G)	X	X	X	X	Objective	Interval	Development Editors	2	2	2	2.0
SPI	X	X	X	X	Objective	Ratio	Project Mngt Software	1	2	1	1.3
CPI	X	X	X	X	Objective	Ratio	Project Mngt Software	1	2	1	1.3
Cpk	X	X	X	X	Objective	Ratio	NA. Manual Calculation	1	2	2	1.7
DD	X	X	X	X	Objective	Ratio	NA. Manual Calculation	1	2	2	1.7
DRE	X	X	X	X	Objective	Ratio	NA. Manual Calculation	1	2	2	1.7

Once the eight measures were selected, a reverse GQM or MQG (Metric-Question-Goal) was carried out to ensure that the measures answer the questions and align to the stakeholder needs. This step is needed because some researchers have argued that the top-down approach ignores what is feasible to measure at the bottom and encourage a bottom-up approach , where organizations measure what is available, regardless of goals [Bache and Neil, 1995; Hetzel, 1993].

The application of the generic GQM framework for the three categories of stakeholders in a software project is as shown below in figure 18.

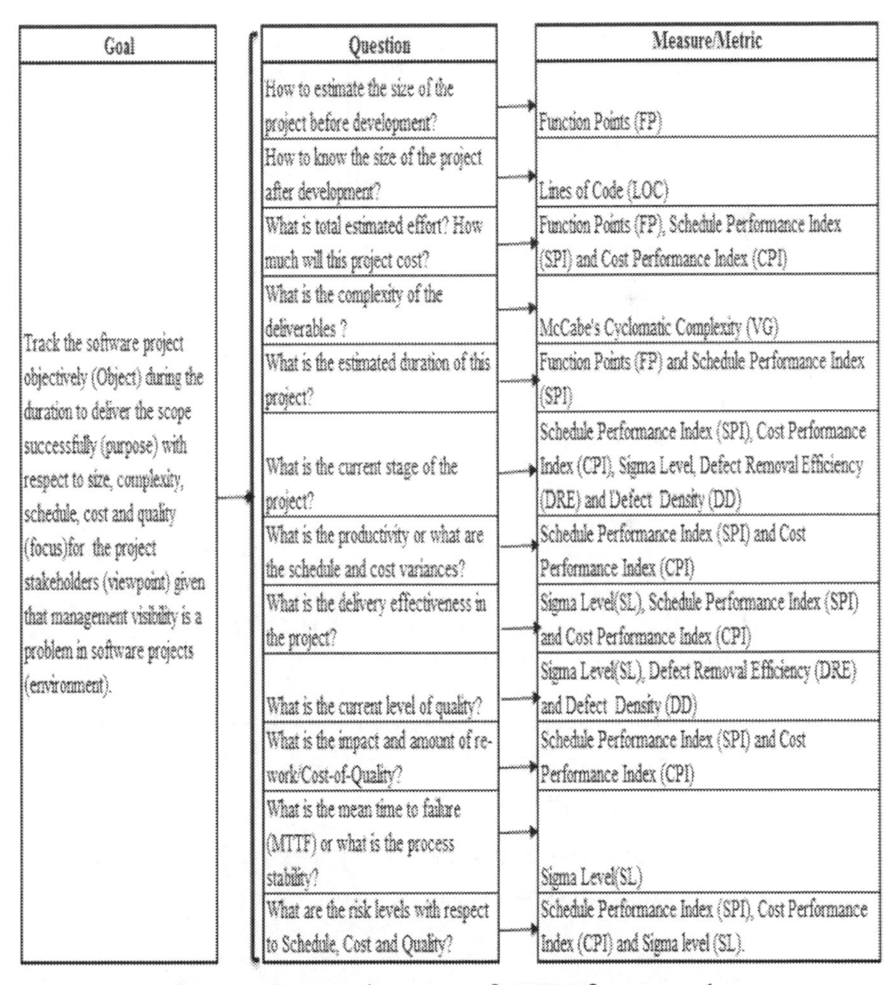

Figure 18: Application of GQM framework

3.7 Hypothesis

A hypothesis is a proposed explanation for an observable phenomenon. It assumes a relationship and puts forward a proposition. The proposition predicts the behavior of variables which must be confirmed by observations made in the real world and can only be confirmed a probably true or false. You can show that the hypothesis has failed, but never prove that it does work. There are two kinds of hypothesis.

1. **Null hypothesis (Ho).**
   The null hypothesis (Ho) typically corresponds to a general or default position i.e. there is no relationship between the measured variables or a potential treatment has no effect.

2. **Alternative hypothesis (Ha).**
   The null hypothesis is typically paired with the alternative hypothesis (Ha), which asserts a particular relationship between the phenomena. Though null and alternate hypothesis are competing statements, the alternative hypothesis need not be a logical negation of the null hypothesis. By providing evidence from observation about the probable truth of one of these hypotheses we can eliminate the other, since both of them cannot be simultaneously true.

Hypothesis testing works by collecting data and measuring how probable the data are, assuming the null hypothesis is true. If the data are very improbable (usually defined as observed less than 5% of the time), then we conclude that the null hypothesis is false and the alternate hypothesis is accepted. In this backdrop, the hypotheses for this research are:

- **The eight core measures (LOC, FP, v (G), SPI, CPI, Cpk, DD and DRE) are the best measures to derive the objective status of their respective attributes.**

- **The eight measures can serve as a generic core set for an accurate and objective status of a software project.**

The key variables which would serve as the operational definition in the hypothesis are:

- **Respective Attributes.**

  LOC and FP are concerned with the size attribute, v (G) for the complexity attribute, SPI for schedule, CPI for cost/budget, and Cpk, DD and DRE deal with quality attributes.

- **Accurate and Objective Status.**

  1. **For Initiators and Implementers**
     +/- 10% of the cost and duration between the baselined project plan and the final cost and duration. Watts Humphrey advocates that a project can be considered successful if is +/- 10% of the initial estimated cost and duration for the agreed scope [Humphrey, 2005].

  2. **For Beneficiaries**
     To address the success criteria of the beneficiaries, we use two criteria.

     a. **Defect rate according to CMMI levels.**
     A defect rate shows how many errors can exist in the production system. According to Capers Jones, the effective defect rates per function point according to the SEI CMMI level in the project are as follows [Kan, 2003]:

     - CMMI Level 1(Initial) – 0.75
     - CMMI Level 2 (Repeatable) – 0.44
     - CMMI Level 3 (Defined) – 0.27
     - CMMI Level 4 (Managed) – 0.14
     - CMMI Level 5 (Optimized) – 0.05

     b. **DRE Levels**
     According to David Longstreet, a software project is mature if the DRE is greater than 45% [Longstreet, 2008].

- **Software Project**

  It can be a Bespoke or a Hybrid project covering all phases of SDLC. Pure COTS implementation projects and software maintenance projects are not in scope as the measures developed are for software development projects.

3.8 Pre-requisites for the Implementation of the Measurement Framework

The prerequisite for implementing the GQM is to understand the critical success factor (CSF) for increased chances of project success. A CSF is something that needs to be in place to achieve the goal. Terry Cooke Davies conducted an empirical research from more than 70 large organizations and identified 12, critical success factors (CSF) essential for project success [Davies, 2002]. The 12 CSFs are:

1. Adequate company-wide education on the concepts of risk management.
2. Maturity of an organization's processes for assigning ownership of risks.
3. Adequacy with which visible risks register is maintained.
4. Adequacy of an up-to-date risk management plan.
5. Adequacy in documentation of organizational responsibilities on the project.
6. Maintain project duration below 3 years if possible (1 year is better).
7. Allow changes to scope only through a mature scope change control process.
8. Maintain the integrity of the performance measurement baseline (PMB).
9. The existence of an effective delivery and management process involving mutual co-operation of project management and line management functions.
10. Portfolio- and programme management practices that allow the enterprise to resource fully a suite of projects that are thoughtfully and dynamically matched to the corporate strategy and business objectives.
11. A suite of project, programme and portfolio metrics that provides direct "line of sight" feedback on current project

performance, and anticipated future success, so that project, portfolio and corporate decisions can be aligned.

12. An effective means of "learning from experience" on projects, that combines explicit knowledge with tacit knowledge in a way that encourages people to learn and to embed that learning into continuous improvement of project management processes and practices.

From the implementation perspective, the GQM provides the measures to track project progress and the conformance to stakeholder goals. In the implementation of the measures, the first and fundamental step is counting the FPs as other measures are derived from it. So a lot of thought and time has to be spent in the initiation and planning phases so that one can get a fairly accurate count of the FPs. This will be extremely challenging considering the fact that requirements will be volatile particularly in the initiation and planning phases. According to SEI, requirements engineering is the most important activity in a software project. Of the eight measures proposed, size (and hence requirements) have a direct bearing on all the measures. Scope constitutes the project's vision, goals, deliverables and boundaries. So when the scope is clear, we get the right information on the measures and derive a clear understanding if there is a significant headway towards the success of the software project.

3.9 Addressing the Limitations of the Proposed Framework

While every project is unique it still shares some generic and common characteristics with other projects. Hence a generic core set of measures to report the software project status should work when a suite of key measures is common, popular and successful in many other industry sectors where decision-making is typically constrained by the lack of information and time. For example:

- In economics, the prosperity index of countries is based on 89 variables grouped into eight categories namely economy, entrepreneurship/opportunity, governance, education, health, safety/security, personal freedom and social capital [http://www.prosperity.com/rankings.aspx].

- In stock markets, there are ten ratios to assess the financial health of the company in three areas namely [Chen and Shimerda, 1981]:
  1. Profitability
     a. Gross Margin Percentage,
     b. Sales, General and Administrative expense %
     c. Operating Margin
     d. Return on Sales
  2. Efficiency
     a. Inventory turn-over
     b. Days Sales Outstanding
     c. Asset Turnover
  3. Capital Management
     a. Current Ratio
     b. Quick Ratio
     c. Leverage Ratio
- In health care, treatment, therapeutic trials and clinical studies on many ailments are conducted using a core set of measures with appropriate indicators.
- The "APGAR" score to assess the health of newborn children is determined using five simple criteria - Appearance, Pulse, Grimace, Activity, and Respiration.
- In sports, the winner of the Olympic decathlon is acknowledged as the world's finest athlete.
- Closer to this subject, there is the six metrics suite "MOOSE" (Metrics for Object-Oriented Software Engineering) proposed by Chidamber and Kemerer. These six metrics are:
  1. Weighted methods per class (WMC)
  2. Depth of inheritance (DIT)
  3. Number of direct sub-classes (NOC)
  4. Coupling between object classes (CBO)
  5. Response set for a class(RFC)
  6. Lack of cohesion in methods (LCOM)

While we can debate the inclusion of more measures to the list of software project measures at additional costs citing different contexts, these core eight measures are meant to be a minimum set capturing critical information most relevant to the stakeholders. **The project reporting can** start with this core set, derive some value and then expand if necessary depending on unique project needs and circumstances. For example, the product manager might want to have more granularities on v (G) for product stability. In that case perhaps cohesion and coupling (between software components) related measures as sub-metrics under v (G) might provide additional information.

3.10 Conclusion

Though lot of research is done on software measures, we still do not have a core set of generic and objective measures for software projects that is validated theoretically and empirically. GQM is found as the most effective measurement methodology for software projects. It is designed and targeted to support the business goals of the organization in an effective and economical way to aid decision making. It facilitates incorporating processes, products and resources thereby making it adaptable to different software project environments.

To control the project for success, the performance needs to be measured with the right measures that are relevant to the stakeholders – initiators, implementers and beneficiaries. Right measures can help to control the project and give the direction needed to correct the project dysfunction given that quantitative methods have proved powerful in other sciences. The eight measures proposed can also be used cost estimation, comparison of products, scheduling of work, measuring productivity to name a few as shown below in figure 19.

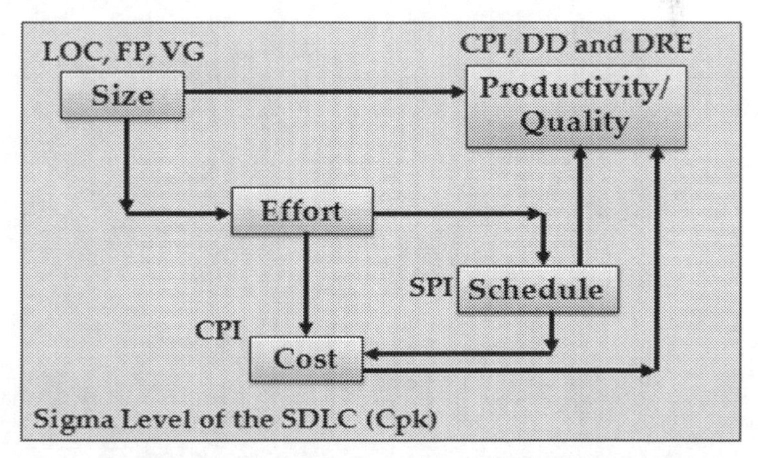

Figure 19: Relationship between 8 Measures

While it is impossible to come up with a "golden set" of measures to suit all types of software projects, the proposed measurement framework with eight measures can be used to capture the key information most relevant to the stakeholders. Depending on specific project situations, more measures can be appropriately added on top of these eight core measures. These eight measures however have to be validated– theoretically and empirically. Chapter 4 covers each of the eight measures in details while the next chapters i.e. 5 and 6 cover the validation of those measures.

Chapter 4: The Eight Core Measures

The proposed generic GQM model for software projects with the eight measures are related to five attributes namely size/scope, schedule, complexity, effort/cost and quality/defects. Anything significantly more might lead to confusion, micromanagement and eventually paralysis not only of the measurement program but also of the software project. One of the reasons for having a small set of core measures is because human beings can cope with 7 ± 2 pieces of information at a time [Miller, 1957] considering that software project management is essentially a cognitive activity. This finding is also supported by other researchers. According to Jim Clemmer, "The key to effective measurement is a small number of simple measures that channel the organization's energy and focus on the strategic areas with the highest potential return" [Clemmer, 1992]. Also Patrik Berander and Per Jönsson recommend that, having too many measurements to collect and analyze might imply that no measurements are used at all [Berander and Jönsson, 2006]. Hence this measurement framework with eight measures should be a good number for consistent and meaningful reporting. Moreover:

- Everald Mills in his SEI document proposes that a software project measurement framework should include at least the four base measures i.e. **size/scope, schedule, effort/cost and quality/defects** [Mills, 1988].
- Steve McConnell identifies **size, productivity (schedule and cost), defects and maintainability** as useful metrics in a software project [McConnell, 1993].
- According to Larry Putnam, the measures underlying effective software management are **effort, time, size, and quality**. [Putnam and Meyers, 2002].
- According to Moller and Paulish, the five widely used measurements are **size, defects, change requests, deviation from schedule and deviation from productivity** [Moller and Paulish, 1993].

- Stephen H. Kan classifies software metrics into three categories: product metrics, process metrics, and project metrics [Kan, 2003].
    1. Product metrics describe the characteristics of the product such as **size, complexity, design features, performance, and quality level**.
    2. Process metrics describe the software development lifecycle process and it is widely accepted that a quality process invariably gives a quality product. Examples include the effectiveness of **defect removal** during development, the pattern of testing **defect arrival**, and the **response time** of the fix process.
    3. Project metrics give more details on the project usually on **schedule, cost and quality**. Examples include the staffing pattern over the lifecycle of the software, cost, schedule, and productivity. Some metrics however might belong to multiple categories. For example, the quality metrics of a project are both process metrics and project metrics.

Below table 22 gives the analysis of the measurement framework with respect to the work done by other researchers.

Table 22: Comparison of Measurement Models

Measure/ Attribute	**Mills**	**McConnell**	**Putnam**	**Moller and Paulish**	**Kan**
LOC/Size	X	X	X	X	X
FP/Size	X	X	X	X	X
V G)/ Complexity or Maintainability		X			X
SPI/ Schedule	X	X	X	X	X
CPI/Cost	X	X	X	X	X
Cpk/Quality	X		X		X
DD/Quality	X		X		
DRE/Quality	X		X		X

All this research suggests that we need to have few relevant and meaningful measures. The eight measures proposed in section 3.6:

1. **Are designed for an objective status based on the stakeholder goals in the project with a clear/objective definition for project success and failure.**

   Objective measures bring certainty and consistency to the measurement framework, no matter which instrument is used, who is measuring and what is being measured and interpreted; within a permissible and predictable range of error.

2. **Go to the next level of detail where measures provide concrete data to track and manage the project progress for taking suitable corrective actions.**

   For instance when Putnam and McConnell talk about size, it is a metric (which is abstract) and not a measure such as LOC or FP (which is concrete).

These two statements (hypothesis) however have to be validated and chapters 5 and 6 cover this. The coming sections talk about each of the eight measures in detail for a good understanding for validating and applying these measures in projects.

4.1 Size/Scope Measures

Quantifying size is one of the most basic and critical activities in software measurement. Size/scope is the work that needs to be accomplished to deliver a product or service with the specified features and functions. Different size measures may be useful at different points in the SDLC as shown in the table 23 below.

Table 23: Different Size Measures

	SDLC Phases			
Size Measure	**Require-ments**	**Design**	**Develop-ment**	**Testing**
Lines of Code		X	X	X
Function Points	X	X	X	X
Documentation Pages	X	X	X	X
Screens and Reports	X	X	X	X
Maintenance Requests	X	X	X	X
Components	X	X	X	
Use Cases	X	X	X	X
Requirements	X	X		
Data stores in System Design	X			
Functions in System Design	X			
Database Size		X	X	
Design Statements		X	X	
Object Points		X	X	
Test Cases				X
Test Procedure Steps				X

Though there are many measures for size as seen in table 23, the two size measures which adhere to the criteria outlined in section 3.6.6 that are most suited for a software project are.

1. Physical size (i.e. Lines of code).

   Physical size is basically the size of the source code of software. Its key features are:
   - Describes the system itself.
   - Represents the size from the development perspective.
   - Easier to define objectively.

2. Functional size (i.e. Function points).

   Functional Size is the software size in terms of the output delivered to the user - the software functionality. Its key features are:
   - Describes the business functionality provided by the system.
   - Represents the functionality from the end user's perspective.
   - It is subjective.

4.1.1 Lines of Code (LOC)

The Lines of code (LOC) measure was formally proposed by Wolverton to formally measure programmer productivity [Wolverton, 1974]. The most common definition of LOC is to count any programming statement without the blank or comment line. It is mainly the count of instruction statements. According to SEI, the LOC is the most widely used metric for program size [Mills, 1988] and Fenton also concurs with this finding [Fenton, 2006]. LOC is very important in the testing phases of the project and maintenance of the product after the project completion particularly for support effort estimation. In addition, LOC is an effective tool for understanding other metrics. Almost any code or testing metric that suffers a sharp spike or sudden drop requires a look at total LOC for more analysis.

There are two major types of LOC measures: Physical LOC (LOC) and Logical LOC (LLOC). Physical LOC is a count of lines in the text of the program's source code including comment lines. Logical LOC measures the number of "statements", but their definitions are tied to specific computer languages. For example, logical LOC in C programming language is the number of key word terminating statements. Consider this snippet of C code as an example of the two types of LOC.

```
For (i = 0; i < 100; i += 1) printf ("Hello World");
/* LOC Determination*/
```

In this example there are:

- 1 Physical Lines of Code (LOC)
- 2 Logical Line of Code (LOC) (for statement and printf statement)
- 1 comment line

Depending on the programmer and/or coding standards, the above program could be even written on many separate lines as shown below.

```
For (i = 0; i < 100; i += 1)
{
   printf("Hello World");
}
```

In this example there are:

- 5 Physical Lines of Code (LOC)
- 2 Logical Line of Code (LLOC)
- 1 comment line

The second example is much easier to read and debug during maintenance than the earlier "spaghetti" code. In addition, it is found to be much easier to create tools that measure physical LOC, and physical LOC definitions are easier to explain. So physical LOC will be used as the definition for LOC.

A study at IBM, on 60 projects of 4000 to 467,000 LOC provided the following equations from regression analysis [Aggarwal and Singh, 2007].

$$\mathbf{E = 5.2 * L^{0.91}}$$
$$\mathbf{DOC = 49 * L^{1.01}}$$
$$\mathbf{D = 4.1 * L^{0.36}}$$
$$\mathbf{D = 2.47 * E^{0.35}}$$
$$\mathbf{S = 0.54 * E^{0.60}}$$

Where,

E = Total effort in person-months
L = Kilo LOC
DOC = Documentation in pages
D = Project duration in months
S = Staff size i.e. number of people.

Although LOC can be useful as it is an easy, fast and inexpensive method to estimate size, it has two important drawbacks.

1. The number of LOC is dependent on the implementation language and coding style of the programmers. For instance, 10,000 LOC in C programming language can be reduced by a couple of hundreds or less LOC in Java, as Java gives many libraries to wrap common programming instructions. (Libraries contain code and data to provide services in a modular fashion to independent programs). Also experienced programmers use less variables and LOC in a program than inexperienced programmers.

2. LOC cannot be determined before coding. Only when coding for the functionality is completed it is possible to count the number of lines in a software product and hence determine the size. Hence the estimate of the size (and subsequent related activities such as effort estimation) cannot be carried out before the development is complete.

4.1.2 Function Points (FPs)

LOC is a technical measure and it does not measure the size of the business functions covered in the project deliverables. In addition, LOC is only known when the development is done. However in many project situations such as effort estimation we have to know the size of the component for scheduling development tasks, budgeting etc. In this situation, functional size measurement is needed as it is aimed at measuring the size of the software product from the perspective of what gets delivered to the end user in terms of business functionality [Albrecht, 1979]. Functional size measurement is considered the ultimate measure of software productivity as it provides the number of functions regardless of the software [Kan, 2003].

There are two primary functional sizing methodologies: Function Point Analysis and COSMIC-FFP and the Function Point Analysis method proposed by defined by Allan Albrecht is the recommended option [Symons, 2001]. FPs provides the measure of software size that can be determined early in the development process and can be very helpful in size/effort estimation. The size reported in terms of FPs is *independent of the computer language, development methodology, technology or capability of the project team used to develop the application.* The high-level procedure for function point counting is as shown below in the figure 20 [IFPUG, 1999, pp 316]

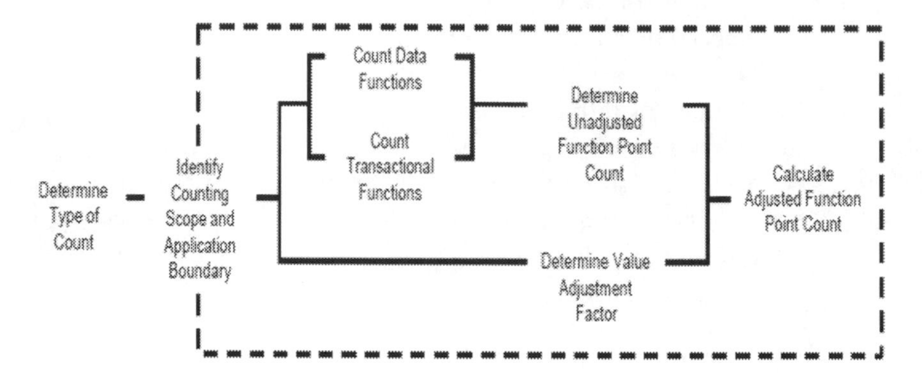

Figure 20: Procedure for Function Point counting

A typical function point analysis consists of the following four steps:

Step 1: Determine the type of function point count (users, purpose, and dependencies)

Based on the design specifications, the following system functions are measured i.e. counted.

- Inputs
- Outputs
- Files
- Inquires
- Interfaces

The three components i.e. External Inputs (EI), External Outputs (EO) and External Inquiries (EQ) add, modify, delete, retrieve or processes information contained in the files and hence are called transactions. The other two components are the system's files viz., Internal Logical Files (ILF) and External Interface Files (EIF) and these are called data functions.

Step 2: Calculate the Unadjusted Function Point (UFP)

These function-point counts are then weighed (multiplied) by their degree of complexity. This is reflected in table 24 based on historical data from real world projects [IFPUG, 1999].

Table 24: Function Point Factors

	Simple	Average	Complex
Inputs	3	4	6
Outputs	4	5	7
Files	7	10	15
Inquires	3	4	6
Interfaces	5	7	10

For example, if an application has nine external inputs which are all simple, eight external outputs which are of average complexity , fourteen simple internal files , seven simple external inquiries and fifteen high complexity interfaces the unadjusted Function Count (FC) = 3*9 +5* 8 + 14*7+ 7*3 +15*10 = 336

Step 3: Determine the Value Adjustment Factor (VAF) to account for factors that affect the project and/or system as a whole.

There will be a number of technical and operational factors a.k.a non-functional requirements (NFR) such as performance, reusability, usability etc that affect the size of the project. From the FP calculation perspective, there are 14 'General System Characteristics", or GSCs, and each one is ranked from "0"for no influence to "5" for essential to account for the complexity of the software system. The ranking for all these factors are added, to get a Value Adjustment factor (VAF) for the measurement object. When the values of these 14 GSCs were summed up we get the "Total Degree of Influence", or TDI value.

$$\mathbf{VAF = 0.65 + (TDI*0.01)}$$

So the VAF can vary in range from 0.65 (when all GSCs are low i.e. GSC = 0) to 1.35 (when all GSCs are high i.e. GSC = 14*5 = 70). Let us assume, all the values from the fourteen factors add up to 44. So the Value Adjustment Factor (VAF) = 0.65 + 0.01*44 = 1.09

Step 4: Calculate the adjusted function point (FP) count

So the adjusted FP = FC * VAF = 336 * 1.09 = 366. This number is the quantifiable metric on the size of the functional scope/size.

Function points have their own advantages and disadvantages. The advantages are:

1. Uses language meaningful to non-programmers including end users.
2. Measures system primarily from logical perspective independent of the implementation technology.
3. Normalizes data across various projects and organizations.
4. Can be used as a reference for estimating key project parameters [Jones, 2002].

    - Function points to the power of 0.40 = Calendar months in schedule
    - Function points to the power of 1.20 = Number of test cases
    - Function points to the power of 1.25 = Count of software defect potential
    - Function points / 150 = Number of development technical staff

The disadvantages with Function points are:

1. It is a labor-intensive method and expensive for applications greater than 15,000 FPs. This is a serious issue given that many software projects that fail are considerably big in size. To put this in perspective, below table 25 gives the FPs counts for common applications [Jones, 2002].

Table 25: FPs of different Applications

Applications	Approximate Number of FPs
SAP ERP application	**296,574**
Windows Vista	**159,159**
Microsoft Office	**97,165**
Typical Airline Reservation System	**~50, 000**
Skype	**21,202**
Google Search Engine	**18,640**
Linux Operating System	**17,505**

2. It requires significant training and experience to be proficient for FP counting.
3. It is subjective to some extent as the functional complexity weights and degrees of influence are determined by trial and error. Though there is lot of historical data to be leveraged to reduce the variation, FPs still bring some amount to subjectivity.

However during the development, the functional points can be validated and the revised FP count can be used for status reporting considering that estimates improve as project progresses. However the FP counts before development can be used to estimate the size for effort estimation.

An important message here is that size (LOC or FPs) and effort generally do not follow the rules of mathematics. This is important when we are analyzing size measurement data for predicting future project events. According to Steve McConnell, *a 10,000 LOC system would require 13.5 staff months to complete. But, ratio math would say that a 100,000 LOC system would not require 135 staff months. But it actually takes 170 staff months* [McConnell, 2006]. *This is illustrated in figure 21*[McConnell, 2006].

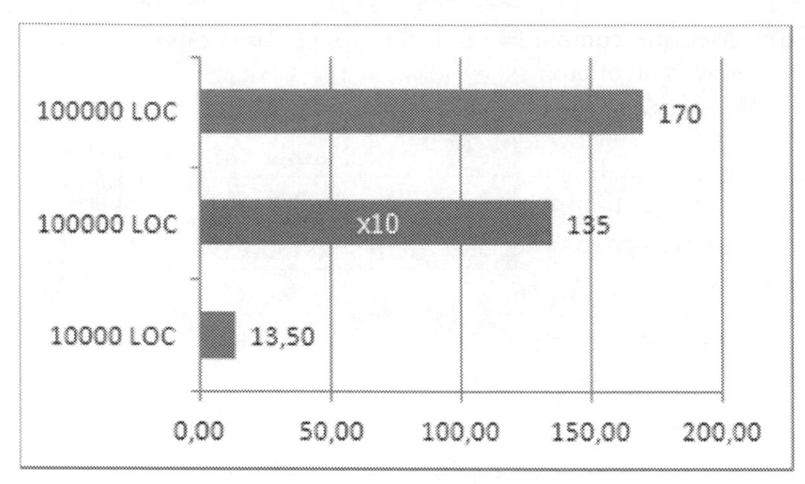

Figure 21: LOC v/s Duration

4.2 Complexity Measure

As mentioned in section 1.6.2, there are four types of project complexity namely: structural complexity, technical complexity, directional complexity and temporal complexity [Remington and Oollack, 2008]. In the backdrop, complexity in software projects is measured either by measuring the product complexity (technical complexity) or by measuring characteristics of the software process (structural, directional and temporal complexity) [Fitsilis, 2009]. This research primarily looks at product complexity.

The McCabe cyclomatic complexity v (G) and Halstead's software science (HSS) are two common code or product complexity measures. However McCabe cyclomatic complexity v (G) is more popular from the implementation perspective in the industry [Fenton, 2006]. In addition (G) is available in most commercial development editors.

To measure the technical i.e. the structural complexity of a program, Thomas J. McCabe developed a measure called the McCabe cyclomatic complexity (v (G)). This measure counts the available decision paths in the program thereby placing a numerical value on the complexity [McCabe, 1979]. This brings practicability and simplicity to the software development process as programmers know that module with many IF/THEN statements is hard to debug and understand. In addition, empirical data has provided a high correlation between defect rates and cyclomatic complexity in the programs.

The McCabe complexity has its roots in graph theory. If G is the control flow graph of program P and G has e edges (arcs) and n nodes, then v *(G)* (P) = e-n+2 where v *(G)* (P) is the number of linearly independent paths in G. In the below figure 22, e = 16 n =13. This gives v *(G)* (P) = 5. More simply, if d is the number of decision nodes in G then v (G) (P) = d+1. In the below figure 22, d = 4. Incidentally McCabe proposed v *(G)* (P) < 10 for each program module P.

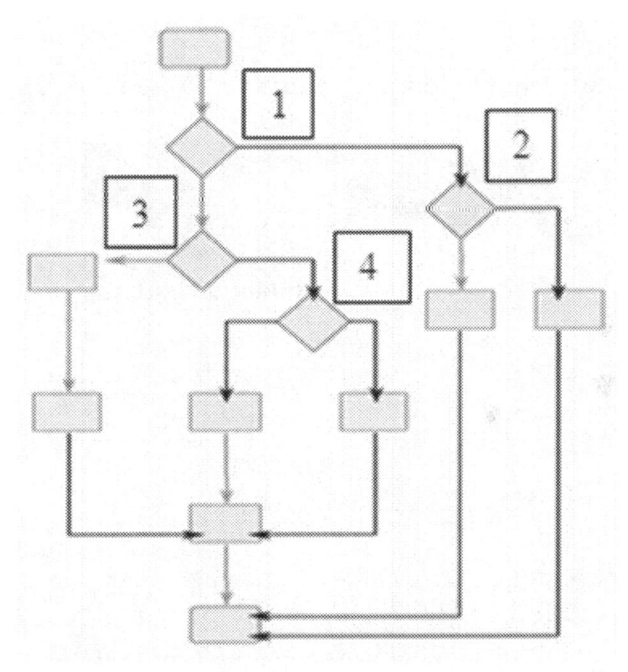

Figure 22: McCabe Complexity Example

For instance, the following simple Java method will yield a McCabe's cyclomatic complexity value of 3.

```
public int getValue(int param1) {
      int value = 0;
      if (param1 == 0)
         { value = 4; }
      else
         { value = 0;}          }
      return value;
      }
```

In the method above, there are two decision points: an `if` and an `else`; and the method's entry point automatically adds one taking the final value to 3. The more control statements such as "if", "while", "for" etc... in the code the higher the McCabe's cyclomatic complexity of the code. Based on the computed value, the below table 26 from the SEI, categorizes the program based on the complexity at class or file level [SEI, 1997].

Table 26: Original McCabe values for Program complexity

Cyclomatic Complexity	Code Complexity
1-10	A simple program, without much risk
11-20	Medium complex, moderate risk
21-50	Complex, high risk
51+	Un-testable, very high risk

So in the above example code, is a "simple program, without much risk" as we the McCabe cyclomatic complexity value of 3.

In addition McCabe's cyclomatic complexity can be leveraged for measuring the business process complexity with tools such as Business Process Modeling Notation (BPMN) and Business Process Execution Language (BPEL) [Fiammante, 2010]. These diagrams coming out these tools look at describing process graphs and their external touch points, which could be software components or human interactions. However as the development complexity is based on the business process complexity, application of **McCabe Cyclomatic Complexity on the source code will determine not just the product complexity but also** business process complexity.

The main advantages of McCabe Cyclomatic Complexity are [Royce, 1998]**:**

i. It is intuitive and easy to apply.
ii. It can be computed relatively early in lifecycle.
iii. Development editors such as Visual Studio integrated development environment (IDE), Eclipse, IntelliJ etc already have McCabe's cyclomatic complexity as part of their metric suite for application lifecycle management.

Some of the drawbacks of McCabe Cyclomatic Complexity are:

i. It is only a measure of the program's technical complexity and not the data complexity associated with the program.
ii. Same weight is placed on nested and non-nested loops. Although the number of control paths relates to code complexity, this number is only part of the complexity picture. According to McCabe, a 3,000 LOC with five IF/THEN statements is less complex than a 200 LOC with six IF/THEN statements [McCabe, 1979].
iii. **McCabe's Cyclomatic Complexity** is a summary index of binary decisions. It does not distinguish different kinds of control flow complexity such as loops versus IF-THEN-ELSES or cases versus IF-THEN-ELSES.

4.3 Earned Value Management (EVM) to measure Project Schedule and Cost

4.3.1 Introduction

Earned value management (EVM) is a project management technique for measuring project progress in an objective manner by combining measurements of scope, schedule, and cost in a single integrated system. It is a systematic project management process used to find variances in projects based on the comparison of work performed and work planned leveraging the fundamental principle that pattern and trends of the past can be good predictors of the future. According to the Project Management Institute

(PMI), EVM is the most effective project measurement tool where schedule and cost can be measured objectively by comparing the amount of work that was planned with what was actually accomplished [PMI, 2004].

The key component of EVM is the project baseline a.k.a. Performance Measurement Baseline (PMB). The **PMB** is the time-phased budget plan for accomplishing work, against which project performance is measured as per stakeholder needs. For management, it provides the ability to review the "original" i.e. baselined projected spending over time and predict a completion date and cost based on a projection of trends experienced to date on the project. For project managers, it provides a tool to review the cost and schedule variance. For team members, it affords an objective perspective on their targets over time.

4.3.2 Building Blocks of EVM

Essentially EVM is about managing a project with a resource loaded schedule. There are three basic elements of EVM which is normally expressed in monetary units such as Dollars, Euros etc. The figure 23 below provides an overview of these three elements [Wilkens, 1999].

1. **Planned Value (PV)**

   PV is the total cost of the work scheduled/planned as of a reporting date. It is recorded when the work is planned showing the cumulative resources budgeted across the project schedule [PMI, 2004].For example, PV can be calculated as Hourly Rate * Total Hours Planned in the project schedule.

2. **Actual Cost (AC)**

   AC is the total cost taken to complete the work as of a reporting date. AC is the indication of the level of resources expended to achieve the actual work performed to date [PMI, 2004].This can be calculated as Hourly Rate * Total Hours Spent.

3. **Earned Value (EV)**

   EV is the total cost of the work completed as of a reporting date [PMI, 2004]. For instance, EV is calculated as Planned Value * % Complete Actual.

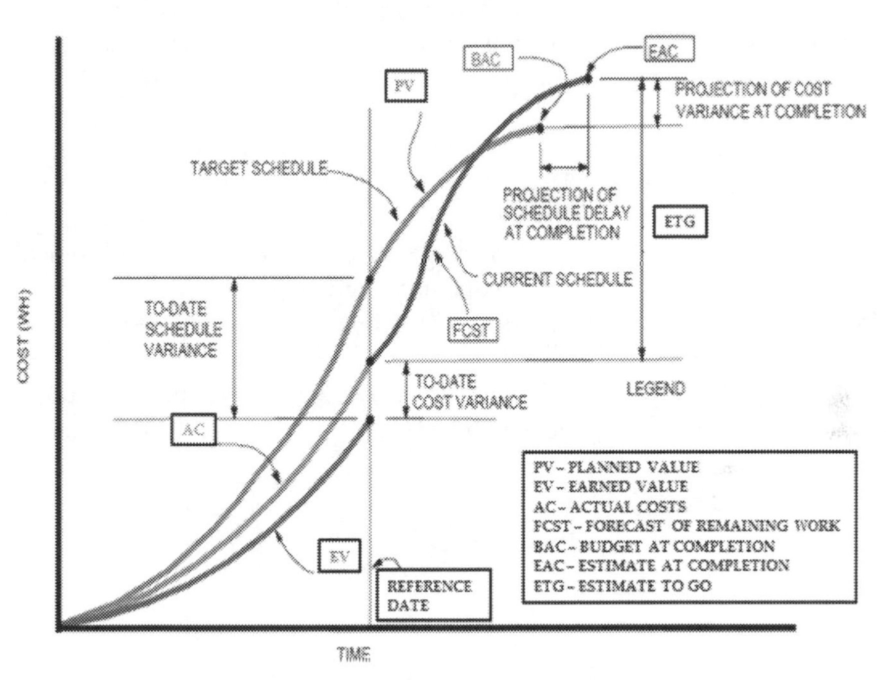

Figure 23: Building blocks of EVM

Using these three numbers i.e. PV, EV and AC, the schedule (SPI) and the cost measures (CPI) are derived.

4.3.3 Implementing EVM

The success of EVM is dependent on the rigorous use of project management tools and processes such as planning, controlling, reporting, forecasting etc. In addition, implementation of EVM demands tailoring of the time/effort reporting and accounting system to allow time and cost to be tracked to specific WBS elements in the project. Following are the four key steps involved in the implementation.

Step 1: Define Requirements Breakdown Structure (RBS), Work Breakdown Structure (WBS) and Organization Breakdown Structure (OBS).

Identifying the scope is the first step in project management and EVM is no exception to this. Once the scope is identified, it is must be decomposed at manageable levels for the right level of traceability. This level of traceability is done with the requirement breakdown structure (RBS). Once the RBS is ready, the next step is building the Work breakdown Structure (WBS).

In the WBS the project is decomposed into elements and each element successively decomposed into component elements until the hierarchy of the complete scope of work is developed. The lowest elements in the WBS are grouped to form work packages (WP) combining related control accounts (CAs). A CA is the management control point at which budgets (resource plans) and actual costs are accumulated and compared to derive the earned value. A WP is the work done at the lowest levels of the WBS and has three elements – technical content, schedule and budget required to accomplish one or more CAs. The relationships between work packages and CAs are as shown in the figure 24 below.

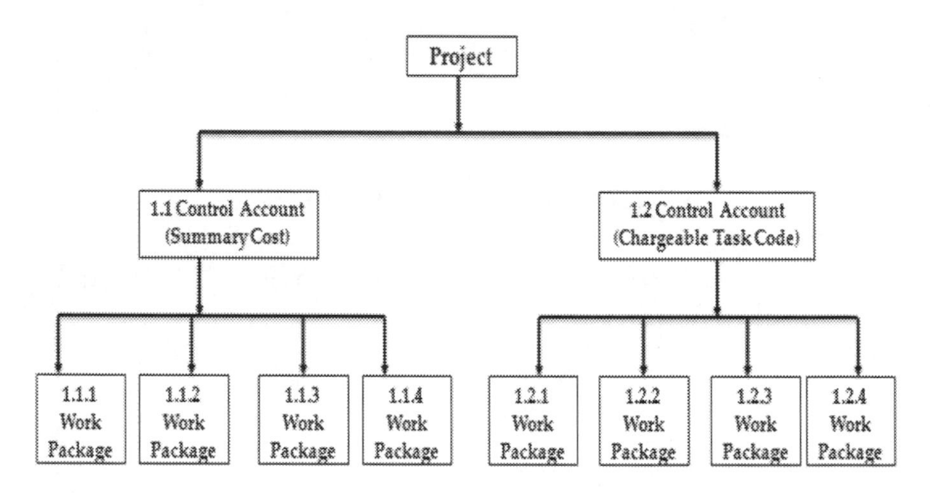

Figure 24: Control Account Structure

Once the RBS and WBS is ready, one must determine who will perform the defined work. Implementation of a software project involves building the right work breakdown structure (WBS) to handle the work/ tasks, scope statement (or requirement breakdown structure (RBS)) to handle the scope and organizations breakdown structure (OBS) with the responsibilities for project control. The integration is done with the CA as it represents the work assigned to one responsible Organization Breakdown Structure (OBS) element to one WBS element. OBS is a hierarchical model describing the responsibilities for project management such as cost reporting, billing, budgeting and project control.

Each defined element in the RBS must be mapped to an appropriate WBS element and every WBS element must have a resource identified from the OBS to complete the specified work. The assignment is a three dimensional plot as shown in figure 25 where each deliverable at every level is a function of WBS, RBS and OBS elements.

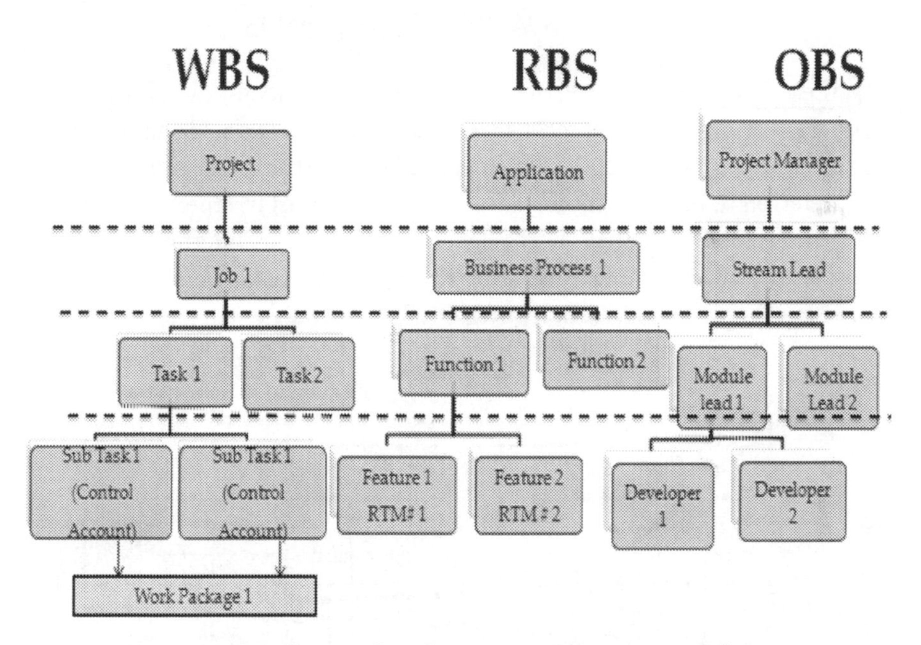

Figure 25: Relationship between WBS, RBS and OBS

Step 2: Schedule the activities, allocate efforts/costs and baseline the plan.

Some projects follow the start-up sequence of scope, schedule, and budget while others follow scope, budget, and schedule. But the outcome at the end of this step is to schedule the activities (after the effort is estimated) with the allocated costs to each CA in the WBS. This is the planned value (PV) in dollars.

The key element associated with the PV is effort estimation which is carried out using one or more of the three types of estimation approaches:

- **Expert estimation:**
  Estimates are produced based on judgmental processes of the experts.

- **Formal estimation model:**
  Estimates are based on formula derived from historical data.

- **Combination-based estimation:**
  Estimates are derived on judgmental or formal estimation or combination of estimates from different sources.

The evidence on differences in estimation accuracy of different estimation approaches and models suggest that there is no "best approach" [Shepperd and Kadoda, 2001]. However to improve the estimation accuracy, it is preferred to have a combination of estimates from independent sources with different approaches. If the level of uncertainty in the estimation effort is high, the project team should be open to rebase line the project because uncertainty reduces during the course of the project as seen in figure 26 [McConnell, 2006].

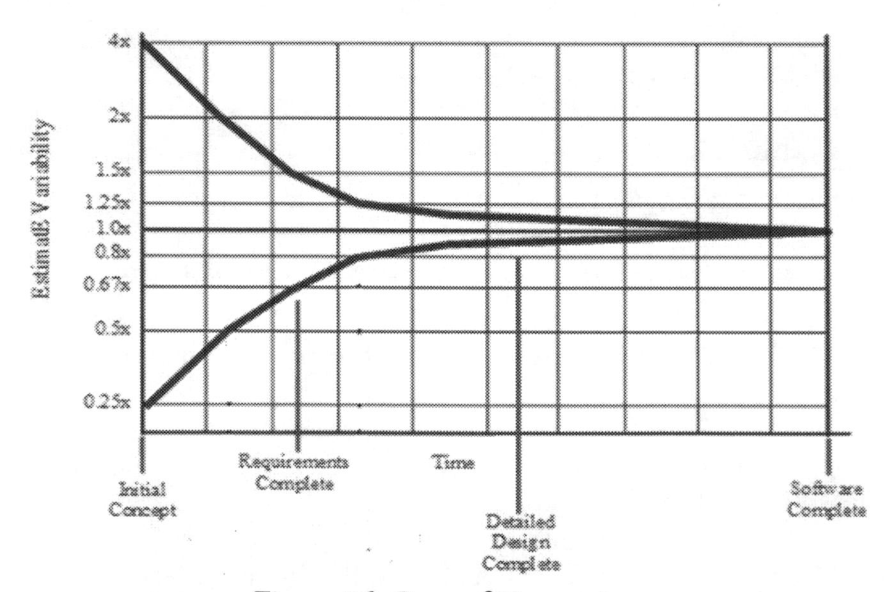

Figure 26: Cone of Uncertainty

The outcome at the end of this step is the spread of the resources over the entire duration of the project i.e. the traditional S-curve also called the PV Curve. This plan should be baselined to form the PMB after the project plan is verified and approved by all project stakeholders.

Step 3: Update project progress regularly to derive EV.

Once the project work is underway, the project plan should be updated regularly on the work accomplished with appropriate earning rules as the percentage of the work accomplished to get the earned value (EV). Earned value (EV) is the key number in EVM. If it is inaccurate then all calculations go wrong. The actual earned value performance measurement will take place within each of the specified CAs. For example, for software projects using the traditional waterfall methodology, one could use the weighted milestone such as the 0-10-35-45-75-90-100 rule to determine the earned value where weighted budget amounts are applied to milestones distributed across duration of task and EV is taken when a particular milestone is reached. In this example, when the project is kicked-off 10% is accomplished, 35% is earned when requirements are elicited, 45% when the design is completed, 75% when coding is done, 90% when testing is complete, and 100% after the application is deployed.

Also actual costs associated with each activity should be simultaneously maintained for each CA and this information can come from time sheets and invoices to the project.

Step 4: Perform calculations, analyze the reports and take corrective actions

To measure schedule, EVM uses a Schedule Performance Index (SPI) which is an index showing the efficiency of the time utilized on the project. SPI can be calculated using the following formula:

SPI = Earned Value (EV) /Planned Value (PV)

Closely associated with SPI is the schedule variance (SV) which indicates how much ahead or behind schedule the project is.

SV = Earned Value (EV) - Planned Value (PV)

To measure cost, EVM uses Cost Performance Index (CPI). According to Fleming and Koppelman, the CPI is likely the single most important metric for any project employing earned value [Fleming and Koppelman, 2000]. CPI is an index showing the efficiency of the utilization of the resources on the project and can be calculated using the following formula:

CPI = Earned Value (EV) /Actual Cost (AC)

Cost Variance (CV) is closely associated with CPI and it indicates how much over or under budget the project is.

CV = Earned Value (EV) - Actual Cost (AC)

Table 27 below shows how the EVM performance measures indicate the project status with respect to the schedule and budget [PMI, 2004].

Table 27: Interpretations of Basic EVM Performance Measures

Performance Measures		SV & SPI		
		>0 & >1.0	=0 & = 1.0	<0 & <1.0
CV & CPI	>0 & >1.0	Ahead of Schedule & Under Budget	On Schedule & Under Budget	Behind Schedule & Under Budget
	=0 & = 1.0	Ahead of Schedule & On Budget	On Schedule & On Budget	Behind Schedule & On Budget
	<0 & <1.0	Ahead of Schedule & Over Budget	On Schedule & Over Budget	Behind Schedule & Over Budget

EVM serves to illustrate the difference between what was planned and what is actually happening against the PMB. It is an early warning system that alerts management to the realities of project performance promoting the philosophy of management by exception.

4.3.4 Limitations of EVM in Software Projects

Though EVM gives an objective way of assessing project performance in terms of schedule and cost, it does have its own limitations.

- EVM only covers cost and schedules and no quality is factored in it. So while EVM indicates a project is under budget, ahead of schedule and scope fully executed, it still has unhappy clients and ultimately unsuccessful projects if the deliverables do not meet the quality expectations.
- EVM requires quantification of a project plan for deriving the PMB. This is often a challenge in software projects as it may be impossible to plan and estimate certain tasks of the projects well in advance, when the project uncovers some tasks and actively eliminates some others during subsequent phases.
- Reliance on a key assumption, that future performance can be predicted based on past performance. Unfortunately there is no guarantee that this assumption will be true when change is inevitable software projects. Even Flemming and Koppleman advocate that EVM can predict and control costs 15- 20% into the project even if the schedule and cost are tracked rigorously [Flemming and Koppleman, 2000].

4.4 Quality Measures

According to Phil Crosby, "Quality is conformance to requirements; both functional and non-functional requirements" [Crossby, 1995]. So when the requirements are met, quality is achieved; and any non conformance to the requirement is reported as a defect. So quality can be improved in the project by identifying and resolving the defects. According to Tom DeMarco, the defect count is the one measure in the project that is always worth collecting [DeMarco, 1995]. In this backdrop, as mentioned in the section 3.2, three measures associated with defects were identified.

1. **Sigma Level (Cpk).**
   This gives an indicator of the effectiveness and stability of the software development process in the project.

2. **Defect Density (DD).**
   This reflects the complexity and stability of different components in the project.

3. **Defects Removal Efficiency (DRE).**
   This provides information on the rate at which defects are resolved.

4.4.1 Sigma Level (Cpk)

Process stability is considered as the core of process management where the process is defined as unique combination of tools, materials, methods, and people engaged in producing a measurable output. When a process is stable and conforming to requirements, it is termed capable. The Process capability is the capability of a process to meet its intended purpose i.e. the ability of a process to produce output within specification limits. When a process is under statistical control, the variation is within predictable limits. A good capable process i.e. the voice of the process (VOP) is one where almost all the measurements fall inside the specification limits which would be the voice of the customer(VOC). Process capability depends on both the stability of the process (VOP) and its ability to conform to customer requirements (VOC). Thus a capable process is a stable process whose performance satisfies customer requirements. This is represented pictorially as shown in the figure 27 below.

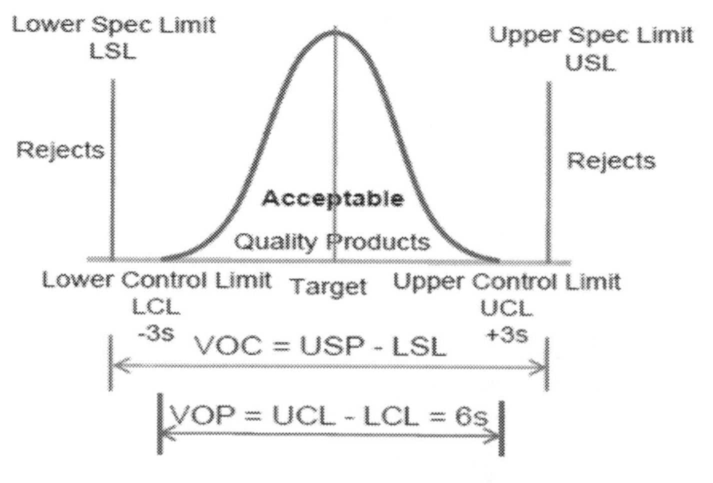

Figure 27: Process Capability

Assuming that the population of data values is normally distributed, there are three statistics that can be used to measure the capability of a process: C_p, C_{pk}, C_{pm}. Cp and Cpk are known as the first generation capability indexes and Cpm is known as the second generation capability index. Though their design principle is approximately the same, they differ from one another by calculation method, by properties and by intended use.

If μ and σ are the mean and standard deviation, respectively, of the normal data and USL, LSL and T are the upper and lower specification limits and the target value, respectively, then the population process capability indices are defined as follows:

$$C_p = \frac{USL - LSL}{6\sigma}$$

$$C_{pk} = \min\left[\frac{USL - \mu}{3\sigma}, \frac{\mu - LSL}{3\sigma}\right]$$

$$C_{pm} = \frac{USL - LSL}{6\sqrt{\sigma^2 + (\mu - T)^2}}$$

- **Process capability index *C*p**

  Process capability index *C*p is a simple relative number comparing value of the required process variability (required tolerance interval) to natural process variability (natural tolerance interval). Capability index *C*p expresses only the potential process capability. It does not represent the position of the natural tolerance interval considering the position of the required tolerance interval. Therefore it does not give a clear answer whether measured value of the searched quality indicator fits within the tolerance interval.

- **Process Capability Index (Cpk)**

  The Process Capability Index (Cpk) is the measurement used in Six Sigma and is the one identified in the measurement framework. The Cpk is a reflection of how well the process occurs within the upper and lower control limits. The higher the value of Cpk the more in-control is the process. It is the limits one would expect virtually all of the occurrences of a process to fall within.

- **Process Capability Index (Cpm)**

  Cpm is rising from new approach to the quality improvement (Taguchi approach). It is not enough to know that measurements are so called good (within the tolerance interval) but more important is knowledge on how good they are. Practical experiences tell that measurements performed directly on products or processes falling outside the tolerance limit (T) are not satisfactory. Process capability index *C*pm defined by Taguchi is intended for decreasing the variability around the target value. Such index enables to determine whether the values of the searched quality index approach to the tolerance limits even when all measurement results fit within the tolerance [Kureková, 2001].

Though Capability index *C*pm represents best the real measurement process capability the Process Capability Index (Cpk) for calculating the sigma level is more widely accepted [Kureková, 2001]. Cpk i.e. Sigma level uses Defects per million opportunities (DPMO) to indicate the effectiveness of the SDLC process; higher sigma level indicates that the process is less likely to create defects [Pyzdek, 2000].

$$\text{DPMO} = \frac{\text{Total Number of Defects}}{\text{Total Opportunity (TO)}} \times 1{,}000{,}000$$

In the DPMO equation, Total Opportunities (TO) is the count of number of defects that can be identified. The number of opportunities is normally proportional to the size (LOC/FPs). DPMO can be then converted to sigma values using Yield to Sigma Conversion table based on the areas under the normal curve shown in figure 28 below.

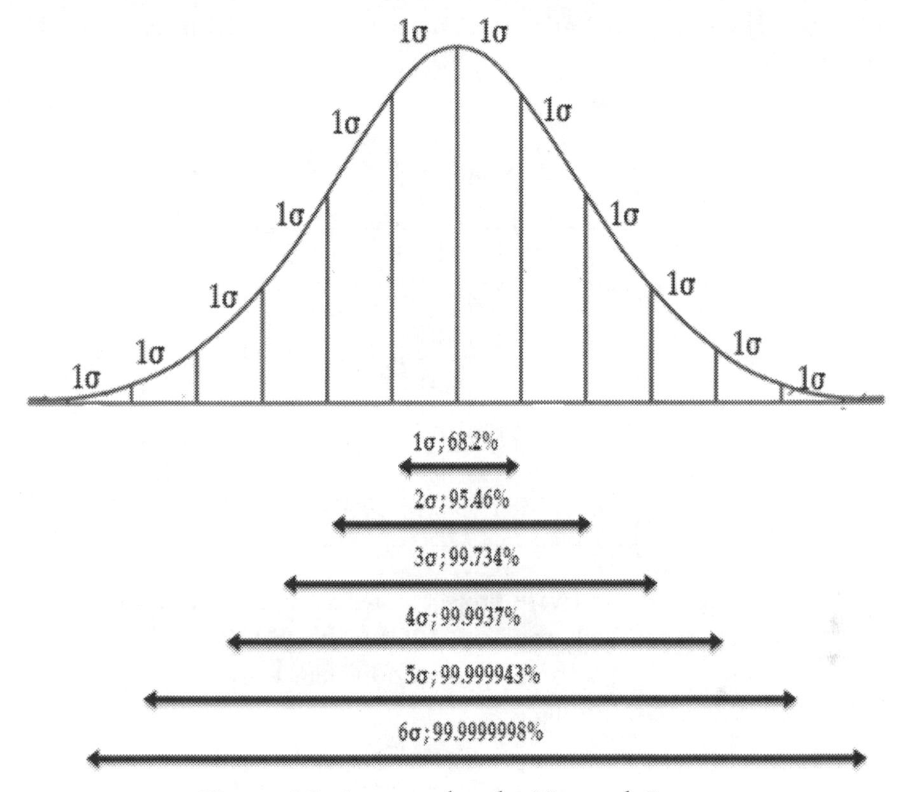

Figure 28: Area under the Normal Curve

As shown in the above figure 28, the area under +/- one standard deviation (sigma) from the mean is 68.2%. The area under +/- two standard deviation is 95.46% %, and so forth. The area under +/- six sigma is 99.9999998% and the area outside the six sigma area is 100% - 99.9999998% = 0.0000002%. If we take the area within the six sigma limit as the percentage of defect-free parts and the area outside the limit as the percentage of defective parts, we find that six sigma is equal to 2 defectives per billion parts of 0.002 defective parts per-million (ppm). This value of 0.002 ppm is from the statistical normal distribution and is very high considering that process shifts and drifts always result in variation in process execution. Research indicates that the maximum process shift is 1.5 sigma. Hence if we account for 1.5 sigma shift in the process, we get a value of 3.4 ppm. This six sigma definition accounting for 1.5 sigma shift has become the industry standard [Kan, 2003]. According to the Yield to Sigma Conversion table, with 1.5 sigma shift, 3.4 DPMO is equivalent to

6 Sigma. The sigma level and the corresponding DPMO is as shown in the table 28 below.

Table 28: Sigma Level v/s DPMO

Sigma Level	DPMO (with 1.5 Sigma shift)
1.5	500,000
3.0	66,800
4.0	6,210
5.0	230
6.0	3.4

For example, in a software project potentially every test case is an opportunity for a defect. If there are 106 test cases/opportunities and 7 defects reported during testing. This is equivalent to 66,037 DPMO (= 7/106 * 1,000,000) or 3 sigma.

4.4.2 Defect Density (DD)

Defect density (DD) is among the most important measures of software reliability as it gives an objective way of comparing different functionalities in the project. It is a measure of the total known defects divided by the size of the software entity being measured. Lower the DD, better it is. According to Steve McConnell, one of the easiest ways to judge whether a program is ready for release is to measure its defect density [McConnell, 1997].

Along with the reliability, defect density is also a reflection of the complexity with respect to size [Hatton, 1996]. The maturity of the development process i.e. the process capability, the skill of the programmers, and the complexity of the development are all reflected significantly in defect density. The formula for Defect density is given by:

$$\textbf{Defect Density (DD)} = \frac{\textbf{Number of Known Defects}}{\textbf{Size of the Software entity}}$$

The "Number of known defects" i.e. the numerator is the count of total defects identified against a particular software entity, in a particular time period in a software project. Size is a normalizer that allows comparisons between different software entities (i.e. modules, releases, products) and is counted either in LOC or FPs. Capers Jones strongly recommends not using LOC as a basis for metrics like DD as they are known to vary by a factor of 10 when compared with FPs. He recommends looking at DD in light of FPs given that FP is invariant with solution size [Jones, 1996]. Hence FPs is considered as the measure for size while calculating DD.

For example, if there are eight defects reported during testing for one function "X" that has nine function points, then the defect density is 8/9 which is 0.9. However a second function "Y" with seven function points might have four defects giving a defect density of 4/7 which is 0.55. So if one has to compare the two functions, function Y is more stable than function X. This information will be particularly helpful as DD serves as a comparator between different modules so that the project team can focus on few critical modules. Also the DD measure can be used to perform a release or iteration wise comparison of quality efforts to see if the quality initiatives implemented are being realized.

4.4.3 Defect Removal Efficiency (DRE)

If a project team has no defect prevention methods in place with no Cpk and DD to indicate the effectiveness, then the project team is totally reliant on defect removal efficiency (DRE). Defect removal is one of the top expenses in any software project and it greatly affects project schedules. Capers Jones says, "Defect-removal efficiency (DRE) is a simple and powerful software quality metric that should be understood by everyone in the software business as it can provide very sophisticated analysis and change "quality" from an ambiguous, amorphous term to a tangible factor [Jones, 1996]". Forrester proposes that, DRE should be the key quality metric for software applications [Forrester, 2003].

The underlying principle behind DRE is that the later the bug is discovered in the project, the greater harm it does and the more it costs to fix given that most forms of testing average only about 30% to 35% in defect removal efficiency levels and seldom top 50% [Jones, 2002]. According to Remus, the relative cost of fixing a defect found in coding, testing and after release are 1:20:82 [Remus, 1983; Kan, 2003]. Hence the DRE metric should be tracked throughout the project cycle. DRE can be of 2 types:

1. DRE across the SDLC
2. DRE in a particular phase.

1. Defect Removal Efficiency (DRE) across the SDLC

This calculation is complex as it includes the latency aspect in defects. In this case the Defect Removal Efficiency (DRE) is calculated as:

$$\textbf{DRE} = \frac{\textbf{Defects removed during a phase}}{\textbf{Total Possible Defects including latent defects}}$$

According to Capers Jones's rule of thumb [Jones, 1996], Function points raised to the power of 1.25 predicts the approximate total number of defects in a new software project. Though ideal value of DRE should be 1, world class organizations have DRE greater than 95% and apply approximately the following sequence of at least eight defect removal activities to reach that level of DRE effectiveness [Jones, 1995].

1. Design inspections
2. Code inspections
3. Unit test
4. New function test
5. Regression test
6. Performance test
7. System test
8. External Beta test

Also defect prediction studies are done on LOC and v (G) metrics using regression based "data fitting" models which provide reasonable estimates for the total number of defects *D* which is actually defined as the sum of the defects found during testing and the defects found during two months after release. Gaffney provided the programming language independent relationship between D and LOC (Lines of Code) using the following polynomial equation [Gaffney, 1984; Fenton, 1999]. The LOC in turn is tied to v (G).

$$D = 4.2 + 0.0015(LOC)^{4/3}$$

So the denominator in the DRE i.e. "Total Possible Defects including latent defects" is Maximum (Capers' Rule of Thumb, Gaffney's Equation). The maximum of the two results is taken so that the DRE value is on a conservative side. For instance, if a software project has 100 function points, the maximum number of potential defects i.e. D as per Capers' rule of thumb will be POW (100, 1.25) which will be 316. For the same project if 12500 LOC are written, then D according to Gaffney's Equation will be 439 defects. So the "Total Possible Defects including latent defects" = Maximum (316,439) = 439.

The table 29 below gives the statistics for project teams on the defect removal efficiency based on their software engineering process maturity [Longstreet, 2008].

Table 29: DRE v/s Process Maturity

Activity	**High Maturity**	**Medium Maturity**	**Poor Maturity**
Requirements Reviews	**15%**	**5%**	**0%**
Design Reviews	**30%**	**15%**	**0%**
Code Reviews	**20%**	**10%**	**0%**
Formal Testing	**25%**	**15%**	**15%**
Total Percentage Removed	**90%**	**45%**	**15%**

So if a project team is at a high level of software engineering maturity, there will be still 10% of the defects not fixed at the end of the testing phase that will be "latent" for the next phase. In the above example, the project would still have up to 44 defects (10%) after all defect discovery and resolution efforts.

2. Defect Removal Efficiency (DRE) in a particular phase

In this case, DRE is calculated as a ratio of defects resolved to total number of defects found. For example, suppose that 100 defects were found during the testing stage and 84 defects were resolved by the development team at the moment of measurement. The DRE would be calculated as 84 divided by 100 which is 84%.

DRE across the SDLC i.e. option # 1 is recommended as it gives a holistic status of the project across the SDLC.

The next level of quality is always tough and it takes time, effort and money to reach there. For example, moving to the next sigma level is a balance between quality and cost as shown in the figure 29 below [Harry, 2000].

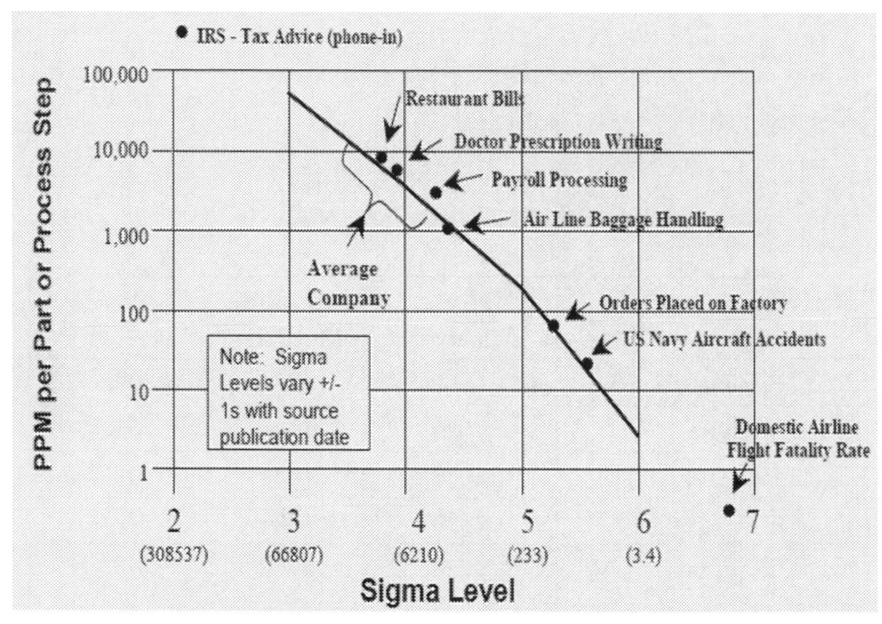

Figure 29: Sigma Level and Business Applications

Depending on the criticality of the application and the cost benefit analysis, we can have permissible acceptance criteria for these three quality measures. For instance, the defects in a software application to update material description in the product master should be at a much higher level of criticality in a pharmaceutical company (due to stringent legislations and technological sophistication for the safety and efficacy of drugs) compared to a manufacturing organization that produces spare parts such as studs, bolts and flanges.

4.5 Conclusion

This chapter covered in detail the eight measures mentioned in section 3.6.6 related to five main attributes namely size/scope, schedule, complexity, effort/cost and quality/defects for a good understanding for validating and applying these measures in projects. This chapter also covered how these measures can be derived. The validation of these eight measures – theoretically and empirically will be covered in the coming chapters i.e. chapter 5 and chapter 6.

Chapter 5: Theoretical Validation of the Framework

5.1 Introduction to Theoretical Validation

According to Lionel Briand, little knowledge exists in the field of software system measurement. Concepts such as complexity and size are very often subject to interpretation and appear to have inconsistent definitions in the literature [Briand, 1994]. As a consequence, there is a need to unambiguously define the measures which can be used to track software projects. One way of doing so is to define precisely the mathematical properties of the measures using measurement theory. Measurement theory is a branch of applied statistics that attempts to:

- Describe, categorize and evaluate the quality of measurements
- Improve the usefulness, accuracy and meaningfulness of measurements
- Propose methods for new and better measurement instruments
- Develop a good intuitive understanding of the concept that is being measured and modeling the intuitive understanding of the attribute in the measure.

From the measurement theory perspective, measurement is a quantification process that is carried out in four steps [Wang, 2003]:

1. **Categorization**.
   Categorization is identification and elicitation of common attributes among a set of interested objects under measurement.

2. **Yard sticking.**
   It is to define the scale of measurement for the identified attributes.

3. **Metrization.**
   Metrization is to find the relationship between the same attribute of two objects or between an attribute and a given measurement scale of the attribute.

4. **Interpretation.**
   Interpretation is to explain the physical or cognitive meaning of the measurement result, measure $\prod = (\alpha, \Omega, \mu)$ where,
   α = Attribute to be measured.
   Ω = Measurement scale
   μ= Unit of measure

According to Briand, theoretical validation determines if the measures follow concepts from measurement theory or axioms. For theoretical validation, measurement theory is a very convenient theoretical framework to explicitly define the underlying theories upon which software engineering measures are based [Braind et al, 1995]. It gives clear definitions of terminology, criteria for experimentation, conditions for validation of measures, foundations of prediction models, empirical properties of measures and criteria for measurement scales. Accordingly, as discussed in chapter 2, the measurement framework will be validated theoretically using the seven theoretical validation criteria proposed by Kitchenham, Fenton and Pfleeger [Kitchenham et al, 1995]. In addition as measures will be part of the measurement framework, we apply the ten questions proposed by Cem Kaner and Walter Bond on the measurement framework to make the transition from theoretical to empirical validation [Kaner and Bond, 2004].

5.2 Overview of the Framework proposed by Kitchenham, Fenton and Pfleeger

Kitchenham et al, categorized software measures into direct and indirect measures [Kitchenham et al, 1995].

- **Direct Measures**

  Direct measures are the building blocks of the measurement frameworks. Direct measurement of an attribute does not involve measurements of other attribute or entity. In the proposed framework, LOC, FP and v (G) are the direct measures as they are directly derived from the product. We consider the characteristics of the domain (i.e., the set of possible attribute instances) and the range (i.e., the possible values to be assigned) for direct measures irrespective of whether the domain and range are finite or infinite, countable or uncountable, bounded or unbounded and whether the mapping is discrete, discontinuous or continuous. The structural model of measurement for direct measures is shown in figure 30 [Kitchenham et al, 1995].

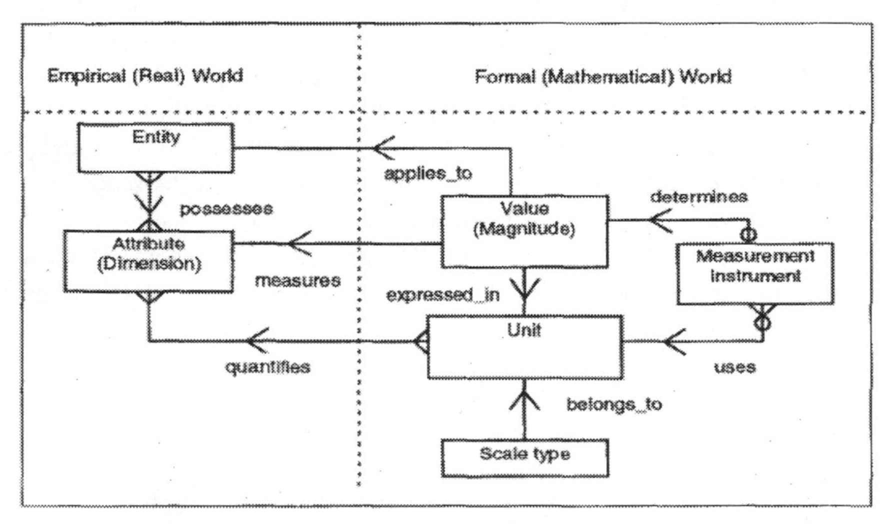

Figure 30: Structural Model for Direct Measures Measurement

- **Indirect Measures**

  However many interesting attributes are best measured indirectly from equations involving other measures. Indirect measurements are often useful in making visible interactions between direct measurements i.e. it is sometimes easier and more informative to see what is happening on a project by using a combination of measures. An indirect measurement is a model that describes how it relates to the attribute it captures and how it needs to be calculated. In the proposed framework, SPI, CPI, Cpk, DD and DRE would constitute the indirect measures. The structural model of measurement for indirect measures is shown in figure 31 [Kitchenham et al, 1995].

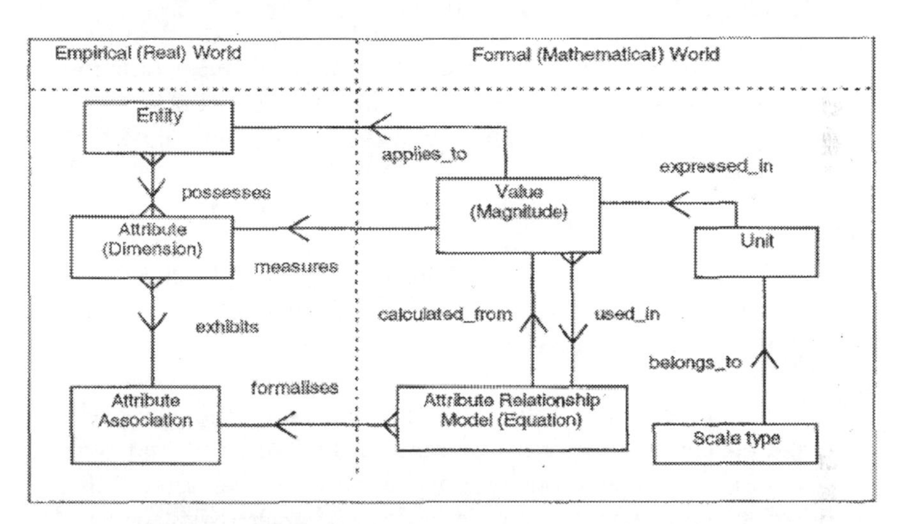

Figure 31: Structural Model for Indirect Measure Measurement

The structural models shown in figures 30 and 31 for direct and indirect measurement have six key elements; some of which were explained at length in section 3.2 of chapter 3. The six elements are:

1. **Entities** are objects we measure. It is the level at which measurement is required and could be products, processes or resources. In the context of this research thesis the entity is the software project.

2. Every entity has **attributes** which describe its characteristics. An attribute may apply to one or more different entities and

different entities may share the same attribute. In the proposed measurement framework, the attributes of the software project (entity) pertaining to process, product and resources are size, complexity/maintenance, schedule, cost and quality.

3. A measure maps an empirical attribute to the formal, mathematical world. A measurement **unit (measure)** determines how we measure an attribute given that an attribute may be measured in one or more units. A unit may also apply to different attributes and different entities. For example, CPI can be used to measure both cost and productivity. Clearly, the specific units are derived with reference to how we want to use our measures. In the context of the proposed measurement framework, units are the eight measures derived from the GQM framework namely LOC, FP, v (G), CPI, SPI, Cpk, DD and DRE.

4. When an attribute is measured, it is on a specific **measurement unit**. To interpret the measured value one has to know:
   i. what entity it applies to
   ii. what attribute is measured and
   iii. what measurement unit the value refers to.

   Also measures need to be defined over a set of permissible values such as finite or infinite, bounded or unbounded, discrete or continuous.

5. Inherent with the measurement unit is the **scale type (or levels of measurement)** namely nominal, ordinal, interval, and ratio as there is a one-to-one relationship between the unit and the scale type. Each scale type has four characteristics namely: distinctiveness, ordering in magnitude, equal intervals and absolute zero. Scale type refers to the relationship among the values that are assigned to the attributes for a variable or unit.

   The scale types can be related to both qualitative and quantitative research. Qualitative research is concerned with measurement on the nominal scale while quantitative research treats measurement on the interval or ratio scale. Knowing the scale type helps in deciding what statistical analysis tool is appropriate. Data originated with categorical scales such as nominal and ordinal use nonparametric statistical tests such as logistic regression models and log-linear models.

Ratio and interval scales which provide continuous data use parametric measures such as t-test, Analysis of Variance (ANOVA), regression, etc. However, most software metrics data comes from a non-normal distribution. This means that non-parametric analysis techniques need to be applied.

6. **Instruments** are used to obtain the measured value of an attribute. Measurement instruments usually detect a single value of an attribute in a particular measurement unit. For example, a parser can be used to measure a direct measure such as LOC for program size. For an indirect measure, an equation can be a potential measurement instrument.

5.3 Applying the Theoretical Validation Criteria

According to Kitchenham et al, both direct and indirect measures are considered valid, if they confirm to the seven criteria namely [Kitchenham et al, 1995]:

1. Scale Validity
2. Appropriate Granularity
3. Representation Condition
4. Unit Validity
5. Protocol Validity
6. Appropriate Continuity
7. Dimensional Consistency

1. Scale Validity

Scales yield numbers that represent properties of the objects they measure. A measure has **scale validity** if it is defined on an explicit, appropriate scale such that all meaningful transformations of the measure are admissible. Each scale type (nominal, ordinal, interval and ratio) denotes a specific set of transformations that dictate how the measure can be used. The scales cannot be changed or transformed that will disrupt the representation. The types of transformations of a scale that are admissible (i.e. maintain correct representation) define the type of scale that have been produced.

Each scale type has specific set of properties involved in the mapping of real world attributes to the numerical world. The scale of one measure however can be transformed into another scale. The mapping from one acceptable (let M be the original measure) measure to another (let M^1 be the new measure) is called admissible transformation [Fenton, 2003]. An admissible transformation is also known as rescaling.

Table 30: Scale Validity

Sl #	Scale Type	Basic Empirical Operation	Admissible Transformations (How M and M^1 must be related)
1	Nominal	Determination of equality/ distinctiveness	1-1 Mapping from M to M^1. The assignment of numbers can be changed as long as the numbers assigned to distinct objects remain distinct.
2	Ordinal	Determination of greater or lesser value	Monotonic increasing function from M to M^1 i.e. $M(x) \geq M(y)$ implies that $M^1(x) \geq M^1(y)$ A monotonic function (or monotone function) is a function which preserves the given order. This transformation does not affect the relative order of the scale values. For example, adding a constant pr multiplying by a positive number.
3	Interval	Determination of equality of intervals or differences	$M^1 = aM + b$ $(a>0)$ Any linear transformation of an interval scale is admissible. However in order to preserve the original ordering of objects, the constant 'a' should be greater than 0.

4	Ratio	Determination of equality of ratios	$M^1 = aM$ (a>0) The only transformation admissible with a ration scale is multiplication by a constant; where the constant should be greater than 0 to preserve the original ordering of the objects.

One scale type is said to be richer than the other if all the relations in the second are contained in the first. At lower levels of measurement, assumptions tend to be less restrictive and data analyses tend to be less sensitive. At each level up the hierarchy, the current level includes all of the qualities of the one below it and adds something new. In general, it is desirable to have a higher level of measurement (e.g., interval or ratio) rather than a lower one (nominal or ordinal). The "Scale Type- Property" relationship is summarized below in table 31 [Fenton and Pfleeger, 1997].

Table 31: Scale Type Property

Sl #	Property	Nominal	Ordinal	Interval (v(G))	Ratio (other 7 Measures)
1	Equality	Yes	Yes	Yes	Yes
2	Uniqueness	Yes	Yes	Yes	Yes
3	Ordinality		Yes	Yes	Yes
4	Difference / Interval ratios			Yes	Yes
5	Value Ratios				Yes

1. **Appropriate Granularity**

Granularity (or the measurement increment) refers to how divisible the value of the measurement unit or measure is. For example, a measurement of an object's weight in grams is more granular than a measurement of the same object's weight in kilograms. A measure has appropriate granularity if the mapping from attribute to measure is not too finely or coarsely grained. The right granularity should reflect the goals of the measurement.

If the granularity is too fine (unless the tool itself provides the fineness), unnecessary effort and cost must be put into data collection, while in the opposite case, the usefulness of the data might be reduced. Too fine a granularity may also make it difficult to provide business value and too coarse a granularity increases complexity and decreases readability.

Of the eight measures in the framework, we categorize seven them as fine (except for FP) as the measurement granularity value is at the lowest level. The Appropriate Granularity validity criterion is summarized in table 32.

Table 32: Granularity of Measures

Measure	Attribute	Granularity
LOC	Physical program size	**Fine**
FP	Functional program size	**Medium**
v(G)	Decision path based program Complexity	**Fine**
CPI	Variation/Adherence to Budget	**Fine**
SPI	Variation/Adherence to Schedule	**Fine**
Cpk	SDLC Process Stability	**Fine**
DD	Module/Program Code Stability	**Fine**
DRE	Quality Achieving Velocity	**Fine**

2. Representation Condition Validity

A measure essentially maps real world attributes to a numerical domain i.e. to a set of integers, rational numbers, or real numbers. In other words, our observations in the real world must be reflected in the numerical values we obtain from the mathematical world. If we call this mapping f, then the measured value of object/entity x is f(x) i.e. "f" is the mapping between real world domain and the numerical domain. This mapping is called representation or homomorphism, because the measure represents the attribute in the numerical world and is illustrated in figure 32 below.

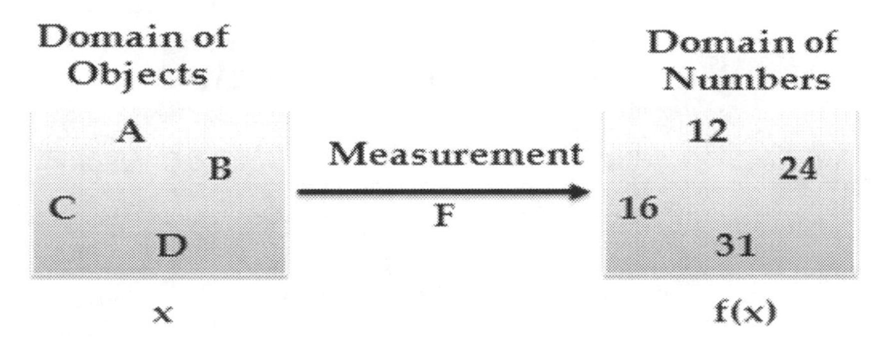

Figure 32: Measurement Mapping

The representation condition asserts that a measurement M must map entities into numbers and empirical relations to numerical relations. For example, the empirical relation "bigger than" is mapped to the numerical relation ">". In particular, we can say that, Program A is bigger than Program B if and only if M (A) > M (B). This statement implies that:

- Whenever Program A is bigger than Program B, then M (Program A) must be a bigger number than M (Program B).
- We can map Program C to a higher number than Program D only if Program C is bigger than Program D.

Representation condition states that the numerical model must always make sense in terms of the real world model it is attempting to describe. Fenton describes the representation condition as a two-way correspondence between a measure and an attribute [Fenton, 1994]. Under the representation condition, any property of the number system must appropriately map to a property of the attribute being measured (and vice versa).

In the proposed measurement framework, LOC satisfies the representation condition for physical program size but it doesn't for functional program size because one can have a poorly written program with more LOC but same or less functionality and vice versa. Hence if the eight measures have to be validated against representation condition, its attributes must be clearly defined in reference to the 12 questions formulated in chapter 3 while deriving the GQM framework. The application of Representation Condition Validity on the eight measures is as shown in table 33 below.

Table 33: Representation Condition (RC) Validity

Measure	Attribute	RC Validity
LOC	Physical program size; size after development	**Yes**
FP	Functional program size; size before development	**Yes**
v(G)	Program Complexity	**Yes**
CPI	Cost/Budget Performance; Productivity	**Yes**
SPI	Schedule Performance	**Yes**
Cpk	Process Stability	**Yes**
DD	Module Stability	**Yes**
DRE	Quality Achieving Velocity	**Yes**

1. **Unit Validity**
   A measure has **unit validity** if the measures used are an appropriate means of measuring the attribute. Although unit definition models may be influenced by a desire to measure a specific attribute of a specific entity type, they usually include some concept of the attribute that is affected by the scale. The eight measures can be categorized as belonging to one of the four types of unit definition models:

2. **Standard Definition**.
   Here measures are selected in reference to a standard definition. For example, LOC is defined as a non-blank, non-comment physical line in the program.

3. **Theoretical Model**.
   This includes referring to the wider accepted theory and involves the way in which an attribute is observed on a particular entity. For example, an "executable statement" can be defined by reference to

the manner in which a compiler handles particular elements of a specific programming language.

4. **Conversion Model.**
   Here the measures are derived from conversion from another measure. For example, "system size" is defined as the sum of its module sizes. This type of definition is controlled by the scale type of the units as the scale type determines the appropriate mappings from one unit to another.

5. **Composite Model**.
   This model is constructed involving several attributes. For example, the unit "hours per line of code" can be used to measure productivity because we define productivity to be the effort to produce a given amount of software.

The application of unit validity on the eight measures is as shown in table 34 below.

Table 34: Unit Validity

Measure	Attribute	Definition Model Type	Unit Validity
LOC	Physical program size i.e. size after development	Standard Definition	**Yes**
FP	Functional program size i.e. size before development	Standard Definition	**Yes**
v(G)	Program Complexity	Standard Definition	**Yes**
CPI	Cost/Budget Performance; Productivity	Theoretical Definition	**Yes**
SPI	Schedule Performance	Theoretical Definition	**Yes**
Cpk	Process Stability	Theoretical Definition	**Yes**

DD	Module Stability	Composite Definition	**Yes**
DRE	Quality Achieving Velocity	Composite Definition	**Yes**

1. Protocol Validity

Protocol validity is whether an acceptable measurement protocol is adopted. For example, in measuring a person's height, the agreed-upon protocol is from the feet to the head, and not including an up-stretched arm. Protocol validity ensures that a specific attribute on the entity is consistent and repeatable. Fundamentally protocol validity of measurement entity is a function of attribute, scales and dimension analysis.

The adherence of the eight measures on Protocol Validity is as shown in table 35.

Table 35: Protocol Validity

	Protocol Validity		
Measure	**Attribute**	**Scale**	**Dimension Analysis**
LOC	Physical program size; size after development	Ratio	Dimensionless
FP	Functional program size; size before development	Ratio	Dimensionless
v(G)	Program Complexity	Interval	Dimensionless
CPI	Cost/Budget Performance; Productivity	Ratio	Dimensionless
SPI	Schedule Performance	Ratio	Dimensionless
Cpk	Process Stability	Ratio	Dimensionless
DD	Module Stability	Ratio	Dimensionless
DRE	Quality Achieving Velocity	Ratio	Dimensionless

Apart from the above five validation criteria, the two remaining criteria (of the seven theoretical validation criteria) namely: Appropriate Continuity and Dimensional Consistency are related to indirect measures [Kitchenham et al, 1995]. An indirect or composite measure is derived by mathematical formula involving other measures.

6. Appropriate Continuity

Valid indirect measures should not exhibit unexpected discontinuities; that is, they should be defined in all reasonable or expected situations. Thus, Measure1 = Count1/Count2, may present problems if Count2 = 0 (when Measure1 becomes infinity) or if Count1 = 0 (when Measure 1 becomes zero). Application of Appropriate continuity is as shown in table 36.

Table 36: Appropriate Continuity

Indirect Measure	Equation	When Numerator is zero	When Denominator is zero
CPI	EV/AC	Earned Value (EV) can be zero if no activity is started against the WBS control accounts with respect to earning rules.	Every activity will consume a resource and a cost comes with every resource. Actual cost (AC) can be zero if there are no costs accounted.
SPI	EV/PV	Earned Value (EV) can be zero before the project starts when no activity is accomplished against the project tasks.	PV can be zero if the effort/cost and schedule is not estimated against the WBS Control Accounts.
Cpk	Yield to Conversion table i.e. Cpk	Cpk is based on DPMO which in turn is dependent on the defects captured against every opportunity.	

DD	Defects Open/FP	This is possible before the "Testing" starts and defects are not captured	This cannot be zero as every development will map to functionality/FPs.
DRE	Defects Resolved/ Total Defects	This is possible when the defects are captured but not resolved.	Every development will have some FPs. We have an empirically based formula to calculate the total defects from FPs.

7. **Dimensional Consistency**

A measure has dimensional consistency if the formulation of multiple measures into a composite measure is performed by a scientifically well-understood mathematical function. Basically this translates to the measures and their attributes agreeing as per dimensional analysis on both sides of the equation for the five indirect measures as shown in table 37.

Table 37: Dimensional Consistency

	Dimensional Consistency		
Measure	**Attribute**	**Scale**	**Dimension Analysis**
CPI	Cost/Budget Performance; Productivity	Ratio	Dimensionless.
SPI	Schedule Performance	Ratio	Dimensionless.
Cpk	Process Stability	Ratio	Dimensionless.
DD	Module Stability	Ratio	Dimensionless.
DRE	Quality Achieving Velocity	Ratio	Dimensionless.

The validation criteria of Kitchenham et al are summarized below in table 38 [Kitchenham et al, 1995].

Table 38: Theoretical Validation Criteria

Sl	Criteria \Attribute	Measures	LOC	FP	V(G)	CPI	SPI	Cpk	DD	DRE
		Measure Type	Direct	Direct	Direct	Indirect (EV/AC)	Indirect (EV/PV)	Indirect	Indirect (Defects Open/FP)	Indirect (Defects Resolved/Total Defects)
		Definition	Physical program size	Functional program size	Decision path based Program Complexity	Variation/Adherence to Budget	Variation/Adherence to Schedule	SDLC Process Stability	Module/Program Code Stability	Quality Achieving Velocity
1	Scale Validity	A measure has scale validity if it is defined on an explicit, appropriate scale such that all meaningful transformations of the measure are admissible. Each scale type (nominal, ordinal, interval and ratio) denotes a specific set of transformations that dictate how the measure can be used.	Ratio Scale	Ratio Scale	Interval Scale	Ratio Scale	Ratio Scale	Ratio Scale	Ratio Scale	Ratio Scale
2	Appropriate Granularity	A measure has appropriate granularity if the mapping from attribute to measure is not too finely or coarsely grained. The right granularity should reflect the goals of the measurement.	Fine	Medium	Fine	Fine	Fine	Fine	Fine	Fine
3	Representation Condition	A measure basically maps real world attributes to a numerical domain i.e. to a set of integers, rational numbers, or real numbers. This mapping is called representation or homomorphism, because the measure represents the attribute in the numerical world.	Yes	Yes	Yes	Yes	Yes	Yes	Yes	Yes
4	Unit Validity	A measure has unit validity if the measures used are an appropriate means of measuring the attribute.	Standard Definition	Standard Definition	Standard Definition	Theoretical Definition	Theoretical Definition	Theoretical Definition	Composite Definition	Composite Definition
5	Protocol Validity	Protocol validity is whether an acceptable measurement protocol is adopted to ensure that a specific attribute on the entity is consistent and repeatable.	Dimensionless	Dimensionless	Dimensionless	Dimensionless	Dimensionless	Dimensionless	Dimensionless	Dimensionless
6	Appropriate Continuity	Valid indirect measures should not exhibit unexpected discontinuities; that is, they should be defined in all reasonable or expected situations. Thus, Measure1 = Count1/Count2, may present problems if Count2 = 0 (when Measure1 becomes infinity) or if Count1 = 0 (when Measure 1 becomes zero)	NA	NA	NA	Numerator i.e. Earned Value (EV) can be zero if no activity is started against the WBS control accounts with respect to earning rules. Actual cost (AC) i.e. denominator can be zero if there are no costs accounted.	Numerator i.e. Earned Value (EV) can be zero if no activity is started against the WBS control accounts with respect to earning rules. Planned Value (PV) i.e. denominator can be zero if the effort/cost and schedule is not estimated against the WBS Control Accounts.	Cpk is based on DPMO which in turn is dependent on the defects captured against every opportunity.	Numerator can be zero before the "Testing" starts and defects are not captured. Denominator cannot be zero as every development will map to functionality/FPs	Numerator can be zero when the defects captured are not resolved. Denominator cannot be zero as every development will have some FPs which in turn can map to the total defects that can be possible in the system.
7	Dimensional Consistency	A metric has dimensional consistency if the formulation of multiple metrics into a composite metric is performed by a scientifically well-understood mathematical function.	NA	NA	NA	Dimensional analysis on both sides of the equation is satisfied.	Dimensional analysis on both sides of the equation is satisfied.	Dimensional analysis on both sides of the equation is satisfied.	Dimensional analysis on both sides of the equation is satisfied.	Dimensional analysis on both sides of the equation is satisfied.

5.4 Validation to Cem Kaner and Walter Bond's Framework.

While the validation criteria of Kitchenham et al [Kitchenham et al, 1995] was at the level of measurement units (or measure), Cem Kaner and Walter Bond proposed ten validation questionnaire at the measurement framework level [Kaner and Bond, 2004]. The objective is to prepare for the implementation of the measures in practical situations.

1. What is the purpose of these eight measures?

All eight measures serve to track the status of a software project (through product, process and resource measures) for providing the right project status to the three groups of stakeholders i.e. initiators, implementers and beneficiaries.

2. What is the scope of these eight measures?

These eight measures should help to analyze projects with the program and modules/work streams within the project so as to make a reasonable comparison to make decisions. While this hypothesis is partially supported by the theoretical validation of the seven criteria explained earlier, the empirical validation of the framework (in chapter 6) will bring in more confidence.

Hypothetically, in a software implementation program for instance, the program manager can use the measurement framework to compare different projects in the program. Depending on the results, scarce resources can be allocated to the struggling area for improvement.

3. What attributes are we trying to measure?

The attributes of the measures are as shown below in table 39 and these are addressed as per the theoretical validation criteria proposed by Kitchenham et al.

Table 39: Attributes of the Measures

Measure	Attribute
LOC	Physical program size; size after development
FP	Functional program size; size before development
v(G)	Program Complexity
CPI	Cost/Budget Performance; Productivity
SPI	Schedule Performance
Cpk	Process Stability
DD	Module Stability
DRE	Quality Achieving Velocity

4. What is the natural scale of the attribute we are trying to measure?

The natural scale of the attribute is closely associated with the definition of the attribute. The scales of the eight measures are as shown in table 40.

Table 40: Natural Scales of the Measures

Measure/Attribute	Natural Scales
LOC/Size	Ratio Scale
FP/Size	Ratio Scale
v(G)/Complexity	Interval Scale
CPI/Cost	Ratio Scale
SPI/Schedule	Ratio Scale
Cpk/Quality	Ratio Scale
DD/Quality	Ratio Scale
DRE/Quality	Ratio Scale

5. What is the natural variability of the attribute?

Measurement is always performed under some degree of uncertainty resulting in some amount of variation in measurement:

- The variability for the size attributes is related to how LOC and FPs are measured. Variation in LOC is low as most of the development frameworks provide this count directly. But the variability in size when FPs is applied is medium because FP calculation is subjective and in the proposed measurement framework it is derived from tables derived empirically from historical data.

- The variability for the complexity measure is low as v (G) is derived directly from development tools.

- The variability for the schedule and cost measures is medium as SPI and CPI calculations depend on the accuracy of the effort estimated and the work accomplished based on the granularity of earning rules.

- The variability of quality attributes depends on how Cpk, DD and DRE are calculated. The variability of Cpk is low as every opportunity is to identify a defect and this can be consistently tracked. However, variability of quality when using DD and DRE is medium as it is based on formulae and tables that are empirically validated. So the variation of the quality measures will be medium.

Therefore the natural variability in all the five attributes depends on how the respective measures are calculated and vary from low (for LOC, v (G) and Cpk) to medium (for FPs, SPI, CPI, DD and DRE).

1. What measuring instrument do we use to perform the measurement?

- LOC can be measured directly with standard development tools. There could even be a parser that strips comments and blank lines.

- FPs will come from a combination of Eclipse plug-in or an equivalent tool and the QSM table [QSM, 2009]. Eclipse is a software development environment comprising an integrated development environment (IDE) and an extensible plug-in system, to check the health of the code.
- Eclipse plug-in tool or an equivalent tool can give the count of v(G).
- Microsoft Project or an equivalent project management software package can provide SPI and CPI.
- The Cpk is basically the opportunity for a defect to occur. DPMO can be then converted to Cpk using "Yield to Sigma" conversion table.
- The DD and DRE can be calculated easily with a normal calculator.

7. What is the natural scale for the metrics?

The scale of the metric can be different from the scale of the underlying attribute. In all the eight measures, we are counting something and this suggests that the scale is either interval or ratio.

- The natural scale for v (G) is interval.
- The natural scale for all other measures is **ratio. For example,** a software component of DD 0.80 is twice as stable compared to a software component of DD 0.40. The same logic applies to LOC, FPs, SPI, CPI, Cpk and DRE.

8. What is the natural variability of reading from this instrument?

Variability might be seen in SPI and CPI as two people might estimate differently and hence different planned values (PV) for a same task. For all other measures, the variability will be low as the readings are coming from either the tables or development tools. This is also reflected in 3.6.6 (table 3.8) where SPI and CPI measures have a FF3 score of 1 as the repeatability and the range of measurement error is high.

9. What is the relationship of the attribute to the metric value?

This is related to construct validity i.e. how do we know the metric measures that attribute. To address the threats on constructs validity, the measures and their objectives were given to software professionals for their feedback through a survey and then implemented in a real world project. This step is covered in detail in the empirical validation with a survey and case studies.

10. What are the natural and foreseeable side effects of using this instrument?

Any implementation of a measurement framework is typically a change management initiative, which is closely tied to cultural and people issues. If people's interests are dependent on the reported metrics, there is a tendency to hide facts or report incorrect data. This means all the eight measures will have some amount of side effects when applied in a software project until an atmosphere of frank and honest status reporting is created.

The Kaner and Bond's ten validation questions are summarized below in table 41 [Kaner and Bond, 2004].

Table 41: Validating Measures using the Ten Questions

Sl #	Question\Measure	Lines of Code (LOC)	Function Points (FP)	McCabe's Cyclomatic Complexity (CC)	Schedule Performance Index (SPI)	Cost Performance Index (CPI)	Sigma Level (SL)	Defect Density (DD)	Defect Removal Efficiency (DRE)
1	What is the purpose of this measure i.e. what is it that we are trying to measure ?	LOC is to measure the Size before development.	FP is to measure the Size after development	CC counts the number of decision paths in the program.	SPI shows the efficiency of the time utilized.	CPI shows the efficiency of the utilization of budget	SL gives the effectiveness of the entire SDLC in the project	DD compares the number of defects in various software components	DRE provides the rate at which defects are resolved
2	What is the scope of this measure?	Size of the work after development	Size of the work before development	Maintenance including Testing	Schedule	Cost or Effort	Quality	Quality	Quaity
3	What attribute are we trying to measure?	Size	Size	Complexity	Duration or Schedule	Cost orEffort	Process Variation	Stability	Speed or Time
4	What is the natural scale of the attribute we are trying to measure?	Ratio	Ratio	Interval	Ratio	Ratio	Ratio	Ratio	Ratio
5	What is the natural variability of the attribute?	Low	Medium	Low	Medium	Medium	Low	Medium	Medium
6	What measuring instrument do we use to perform the measurement?	Tools in the Development Environment	Combination of Eclipse plug-in & QSM Table	Eclipse plug-in	MS project or equivalent	MS project or equivalent	Sigma Conversion Table	Calculator	Calculator
7	What is the natural scale for this metric?	Ratio	Ratio	Interval	Ratio	Ratio	Ratio	Nominal	Nominal
8	What is the natural variability of readings from this instrument?	Low	Low	Low	Medium	Medium	Low	Low	Low
9	What is the relationship of the attribute to the metric value ?	High.	Medium.	High	Medium	Medium	High.	Medium	Medium
10	What are the natural and foreseeable side effects of using this instrument?	Medium	Medium	Medium	Medium	Medium	Medium	Medium	Medium

5.5 Decision Criteria

Implementation of the measurement framework is closely associated with the decision criteria. Decision criteria, or factors, are the specific measures used to determine if any corrective action is needed from the information coming out from the eight measures. Decision criteria essentially help to interpret the measurement results based on:

- Stakeholder requirements
- Industry standards
- Historical data from similar projects
- Future predictions and needs.

The three common decision criteria measures are:

- **Thresholds.**
  Thresholds are established boundaries that when crossed indicate that action is needed. For example, empirical evidence shows that modules with a McCabe's Cyclomatic Complexity greater than 10 at class level are more error prone and harder to maintain [Jones, 1991].

- **Variances.**
  Variances compare actual values with expected values and make decisions based on the magnitude of the difference. If the measurement value for a variance is in the unacceptable range, an analysis must be conducted to determine the cause(s) and corrective action should be taken.

- **Control limits.**
  Statistical process control charts with control limits are one of the classic ways of controlling processes. Control charts, also known as Shewhart charts essentially determine whether a process is in a state of statistical control. To establish control limits, the mean becomes the center line (CL) and zones are created at ±1, ±2 and ±3 standard deviations from the mean. The upper control limit (UCL) is set at plus 3 standard deviations above the mean and the lower control limit (LCL) is set at minus 3 standard deviations below the mean. So if

the value given by the measure is above or below the control limits, suitable corrective action is required.

The three decision criteria can be applied on the eight measures though one particular criterion might work well over others. For instance applying control limits to the sigma level works better than variances. The application of the decision criteria specifically on this research thesis is explained in the hypothesis section in section 3.7. The table 42 below shows the preferred decision criteria on the eight measures.

Table 42: Preferred decision criteria

Measure	Preferred Decision Criteria
LOC	Threshold
FPs	Threshold
v(G)	Threshold
CPI	Threshold/Variances
SPI	Threshold/Variances
Cpk	Control limits
DD	Threshold
DRE	Threshold

5.6 Conclusion

One way of unambiguously defining the measures is to describe the mathematical properties of the measures using measurement theory. Hence the proposed measurement framework is theoretically validated using concepts from measurement theory proposed by Kitchenham, Fenton and Pfleeger. The theoretical validation also minimizes the chances of failure encountered during the implementation of the measurement framework by addressing exceptions and providing evidence as to whether a measure really captures the internal attributes they purport to measure. Hence theoretical validation is a necessary step before empirical validation takes place. To further bolster this validation, the framework is run through the ten questions of Kaner and Bond [Kaner and Bond]. In addition this validation also serves as a link between the theoretical and empirical validations.

Though theoretical validation using measurement theory is getting a great deal of attention from researchers, industry practitioners still rely on empirical evidence of a measure's utility. In the next chapter, the measurement framework is empirically validated with a survey and case studies (controlled and uncontrolled).

Chapter 6: Empirical Validation of the Framework

6.1 Introduction

Software engineering is not just the tools and processes, but also social and cognitive processes involved in it. Within software engineering, empirical software engineering (ESE) is a branch that involves the collection and analysis of data and experience that can be used to characterize, evaluate and reveal relationships between software development deliverables, practices and technologies [Wohlin et al, 2000]. The goal is to observe complex social settings an i.e. context where the interaction among humans is the critical factor that determines the quality and effectiveness of the results being produced. In this scenario, empirical studies are crucial to the evaluation and validation of processes including measurement frameworks.

Empirical validation is basically hypothesis testing using statistical analysis of data obtained via surveys, experiments (controlled and uncontrolled) and case studies to identify and understand the relationships between different variables. Software engineering experiments are often quasi-experiments where it is not possible to select participants in experiments by random. In this backdrop, this chapter addresses the eight empirical validation criteria identified in chapter 2. Six of the eight criteria of empirical validation are proposed by Schneidewind [Schneidewind, 1992]. They are:

- **Association.**
  A measure has association validity if it has a direct, linear statistical correlation with an external quality factor.

- **Consistency.**
  A measure has (rank) consistency if it shares the same ranking as the quality factor.

- **Discriminative power.**
  A measure has discriminative power if it can show a difference between high-quality and low-quality components by examining components, above or below a pre-determined critical value.

- **Tracking.**
  A measure has trackability if the metric changes as the external quality factor changes over time.

- **Predictability.**
  A measure has predictability if it can predict values of an external quality factor with an acceptable level of accuracy.

- **Repeatability.**
  A measure has repeatability if it is empirically valid for multiple different projects or throughout the lifetime of one project.

The remaining two empirical validation criteria are proposed by Kitchenham et al [Kitchenham et al, 1995] and they are:

- **Attribute Validity.**
  A measure has attribute validity if the attribute is exhibited by the entity measured.

- **Instrument Validity.**
  A measure has instrument validity if the measurement instrument is valid and properly calibrated.

Measurement researchers have proposed the empirical validation of the measurement framework to be carried out with [Braind, 1995; Fenton, 2006; Soni et al, 2009]:

1. A survey from seasoned software professionals and
2. A case studies of a real world software project – controlled and uncontrolled.

6.2 Empirical Validation with a Survey

Survey is a method for collecting quantitative information about items in a population. It is a non-experimental, descriptive research method to collect data on phenomena that cannot be directly observed. It is often an investigation performed in retrospect. Survey converts abstract and psychological information such as perceptions, emotions, attitudes etc into hard data for analysis. There are two basic types of surveys:

1. **Cross-sectional surveys.**
   Cross-sectional surveys are focused on finding relationships between variables in a population at a particular point in time.

2. **Longitudinal surveys.**
   Longitudinal surveys gather data over a period of time to find relationships between variables.

In this research, cross sectional survey will be employed as we want to know the response of the industry practitioners on the two hypotheses at a defined point in time.

The unit of analysis or the target of the survey is project stakeholders, who are grouped under three categories namely: initiators, implementers and beneficiaries as explained in chapter 3. The survey was designed on a Likert scale for the population of software industry practitioners'. On every question the respondent was asked to evaluate their level of agreement or disagreement on the applicability of the measure in achieving its intended purpose using the five ordered response levels (namely strongly disagree = 1, disagree = 2, neutral = 3, agree = 4, and strongly agree = 5). The responses were treated as ordinal data as they have an inherent order or sequence and one cannot assume that respondents perceive all pairs of adjacent levels as equidistant. For example, one cannot assume that the difference between "agreeing" and "strongly agreeing" is the same as between "agreeing" and being "neutral".

Surveys generally tend to be weak on validity and strong on reliability [Alreck, P. L and Settle, R. B, 1995]. The artificiality of the survey format puts a strain on validity as the respondents real feelings are hard to grasp in terms of such dichotomies as "agree/disagree," "support/oppose," etc. These are only approximate indicators of what the respondents have in mind. Reliability, on the other hand, is relatively a clearer matter. Survey research presents all subjects with a standardized stimulus, and so goes a long way toward eliminating unreliability in the researcher's observations. Careful wording, format, content, etc. can reduce significantly the subject's own unreliability.

Hence to address the validity threats of the survey questions, two pilot studies (email and interview) were conducted on a small group of six different stakeholders from different companies and countries. The first pilot study was to ensure that the survey questions address the content, criterion, and construct validity objectives. Feedback was collected and the survey questions were deliberated; however no responses were collected. To ensure there was no ambiguity these questions were reworked. Then the second pilot study was conducted on the same group and their responses were collected. Two purposes were accomplished here from the two pilot studies.

1. The first purpose was to check the validity of the survey questions.
2. The second purpose was to start the first phase of the "Test-Retest" reliability test.

When the second pilot study was administered, there was no feedback on the way the questions were constructed indicating un-ambiguity in the survey questionnaire. In addition, some people who earlier had interviews were asked to respond to email and vice versa for criterion validity. The survey questions that were employed are in Appendix 2.

As mentioned before in Chapter 2, the problem with test-retest reliability estimate is the potential for "carry over" effect between the two testings i.e. the respondents might remember the answers they gave the first time and simply repeat them. So to address this issue, after six weeks, the same questionnaire was again sent to the same group for their responses to ensure reliability and repeatability. This was taken as their final response for survey analysis. The Pearson's correlation for all the measures between the second pilot test and the final survey response was close to the recommended threshold value of 0.90 between the data sets of the two tests as shown in the figure 33 below satisfying the **repeatability criteria** for empirical validation [Schneidewind,1992].

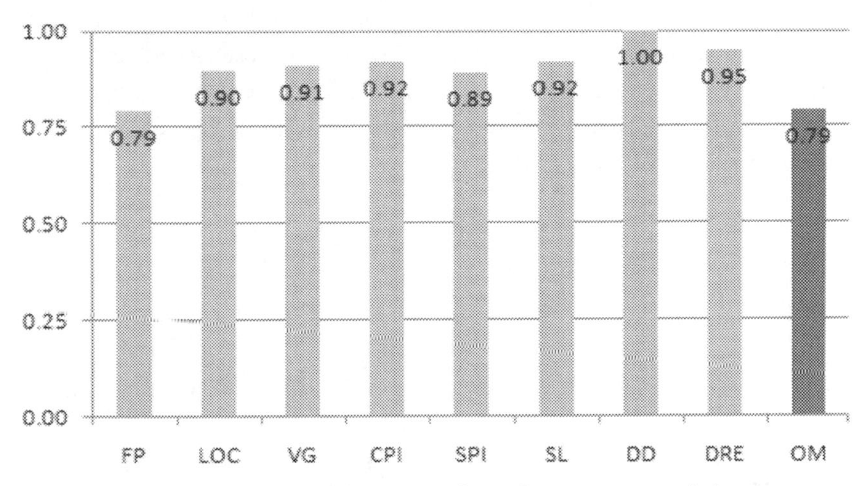

Figure 33: Correlation Values for Survey Validity

As a second test for reliability, the Cronbach's alpha for the same data set was carried out. The Cronbach's alpha is a measure of internal consistency i.e. how closely related a set of items are as a group. A high value of alpha is evidence that the items measure the underlying construct. It is basically an average inter-correlation among the items. The overall value of Cronbach's alpha was found to be 0.78 which again exceeded the threshold of 0.70 set for **repeatability criteria** for reliability/repeatability [Schneidewind, 1992].

After ensuring that the survey questions are valid and reliable, a representative sample from the target population was determined based on factors such as heterogeneity of the population, timelines, necessary degree of precision and other implementation factors. An optimal sample size was required because, if the sample size is too small, important research observation can be lost and if the sample size is too big, one could waste valuable time and resources. Hence sample size determination is typically a balancing act between precision (or reliability) and cost of the survey. In addition, the error coming from the sample is inversely proportional to the square root of the sample size [Polgar, Stephen and Thomas, Shane, 2000, pp 36]. This means that although a sample of 8000 is four times as large as a sample of 2000, it can only be twice as accurate, since the square root of four is two. Hence there is some "optimal number" associated with the sample size.

Since survey research is always based on a sample of the population, the success of the research is a function of the representativeness of the population of concern. A good sample size (SS) also helps in generalizing the findings using statistical inferential techniques to the entire population. The sample size (SS) is given as:

$$\mathbf{SS = Z^{2} * (P) * (1-P) / C^{2}}$$

Where:

- Z = Using a Z-score table, the Z value for 95% **confidence level** is 1.96.
- P= Predict the **proportion** of the study.
- C = **Confidence interval** i.e. margin of error and it is taken as 10%.

To determine the Lower Specification Limit (LSL) of sample size, the participation factor of 0.7 was applied because in the pilot studies conducted, 70% of the participants had responded to the survey. This gave the sample size a LSL value of 81. To determine the Upper Specification Limit (USL), a 90 % participation in the survey was considered, and this provided the USL of 138. So the survey was targeted between 81 to 138 respondents.

The survey was then ultimately rolled out and responses were collected from 110 professionals who shared 1772 years of software project experience. These respondents were from 29 countries from Vietnam in the far-east to Chile in the far-west and included Chief Information officers (CIO), Directors, Program Managers, Project Managers, Business Analysts, Quality Managers, Developers, Architects, Metrics specialists and Users. The organizations of these respondents were at different CMMI levels and were from industry sectors such as Professional services, Telecom, Banking, Insurance, Health care, Pharmaceuticals, Energy and Retail. The profile details of the survey respondents are as shown in the figure 34 below.

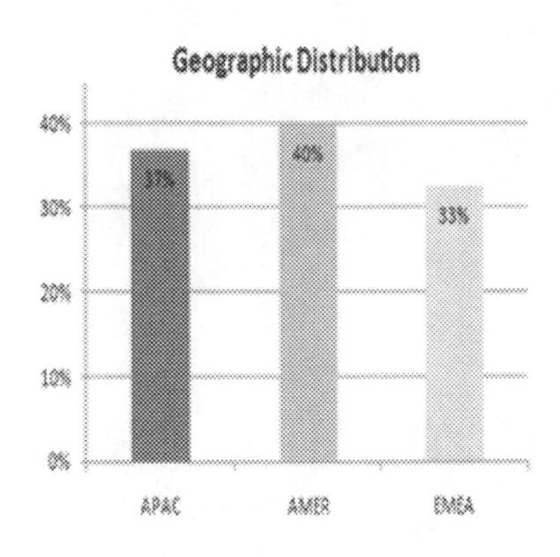

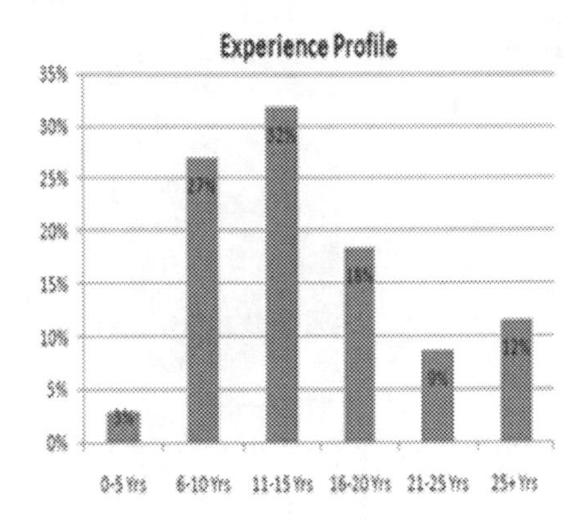

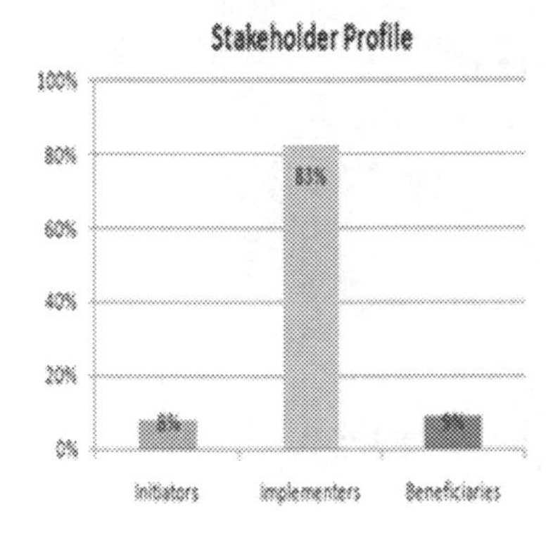

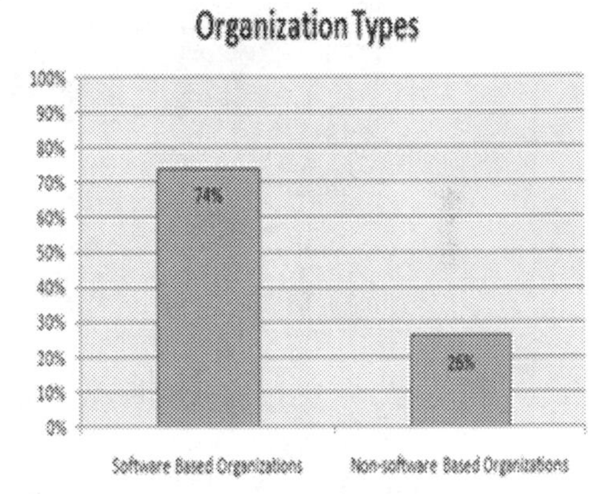

Figure 34: Profile of the Survey Respondents

6.3 Analysis of the Survey

Figure 35 shows how the survey and its statistical analysis were conducted. After the right sample size of the stakeholders is determined, survey analysis is carried out to give voice to that data using descriptive and inferential statistics where the data set is analyzed. Finally the reports are presented to support fact-based decision-making.

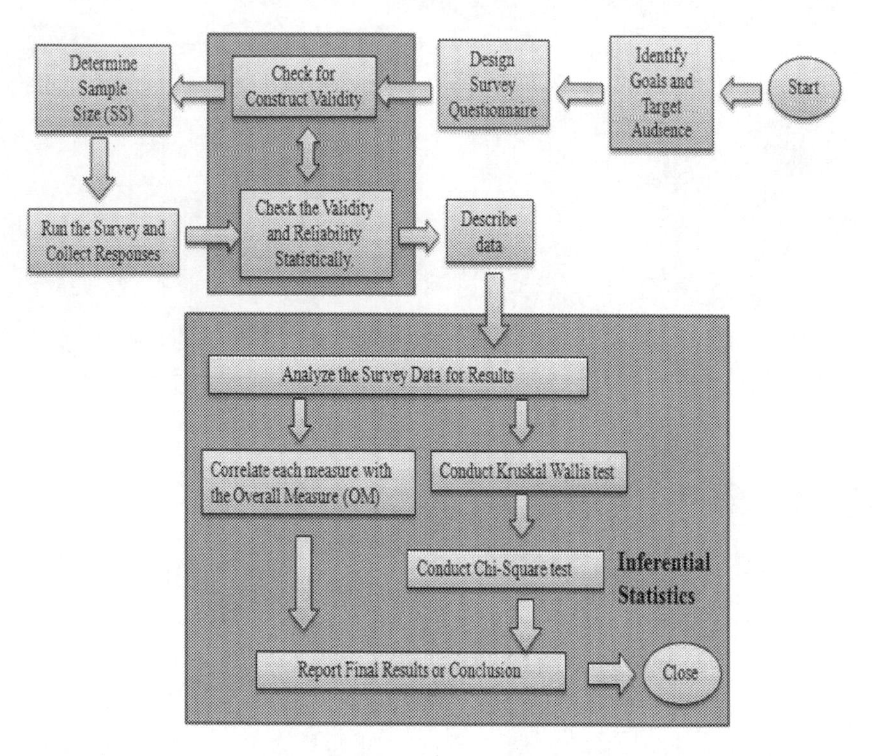

Figure 35: Survey Analysis Flowchart for Ordinal Data

Descriptive statistics summarize a data set quantitatively without employing a probabilistic formulation. They describe data sets that are measures of central tendency and variability. As the survey responses were ordinal data, the measures such as mean, standard deviation and variance were not used in the descriptive statistics (and for analysis) as shown in the table 43 below. The overall measure (OM) is the "holistic"
measure where the 8 Core measures in totality can be used to describe the accurate and objective status of the software project.

Table 43: Descriptive Statistics of the Survey data

Metric	FP	LOC	VG	CPI	SPI	SL	DD	DRE	OM
Median	4.0	3.0	4.0	4.0	4.0	4.0	4.0	4.0	4.0
Mode	4.0	3.0	4.0	4.0	4.0	4.0	4.0	4.0	4.0
Skewness	-1.5	-0.2	-0.8	-1.2	-0.7	-0.3	-0.5	-0.9	-0.9
Kurtosis	4.2	-0.3	0.8	2.8	0.5	-0.4	-0.2	1.3	0.8
Range	4.0	4.0	4.0	4.0	4.0	3.0	3.0	4.0	3.0
1st Quartile	4.0	3.0	3.0	4.0	3.0	3.0	3.0	4.0	3.0
3rd Quartile	4.0	4.0	4.0	4.0	4.0	4.0	4.0	4.0	4.0
Maximun Value	5.0	5.0	5.0	5.0	5.0	5.0	5.0	5.0	5.0

For the measurement error to be small, best possible reliable data should be available. Reliability of the data in the eight measures is compared using the index of variation (IV) and is an indication of how much an observed score can be expected to be the same if observed again. IV is simply a ratio of the standard deviation to the mean. Smaller the IV, more reliable is the measurements. Table 44 below indicates that data for FP and CPI is more reliable than data collected for LOC.

Table 44: Index of Variation

	FP	LOC	VG	CPI	SPI	SL	DD	DRE	Overall
Reliability (Index of Variation)	0.19	0.30	0.23	0.19	0.23	0.22	0.22	0.21	0.20

As the survey responses came in a ranked order, the non-parametric tests (distribution free) were used as they do not depend on normal data. The Kruskal–Wallis (H-value) test was picked for one-way analysis of variance for testing equality of population medians between the two groups i.e. each of the eight measures and the overall measure (OM). The Kruskal–Wallis (KW) procedure tests the null hypothesis that *k* samples from possibly different populations actually originate from similar populations, at least as far as their central tendencies, or medians, are concerned. The test results are shown in table 45 below.

Table 45: Kruskal Wallis (KW) Test Results

Sl #	Test	H-Value	P-Value
1	FP v/s OM	0.434	0.511
2	LOC v/s OM	15.621	0.001
3	VG v/s OM	0.817	0.366
4	CPI v/s OM	1.629	0.202
5	SPI v/s OM	0.038	0.846
6	SL v/s OM	0.087	0.768
7	DD v/s OM	1.781	0.182
8	DRE v/s OM	3.355	0.067

The P-Value (which is adjusted for ties) was greater than 0.05 for all measures except LOC. So the null hypothesis was accepted i.e. there is no difference in the way the respondents rated each of the seven measures (except LOC) with the overall measure. When the null hypothesis is rejected, the result is said to be statistically significant i.e. a result is unlikely to have occurred by chance. This also indicates that there is a correlation between each of the seven measures and the overall measure (OM).

To confirm the results from KW test, the Chi-square test for independence was conducted. The chi-square test is used to test if a sample of data came from a population with a specific distribution. The chi-square test is defined for the hypothesis:

Ho: The data follow a specified distribution.
Ha: The data do not follow the specified distribution

The data sets included the means of the eight measures and the overall measure (OM). The P-value was 0.23 and was greater than 0.05 again implying that null hypothesis should be accepted as there is a correlation i.e. the data sets follow a specified distribution between the two data sets.

The survey data was further analyzed with correlation and regression analysis to understand how the dependent and the independent variables are associated. There was a fair degree of correlation between the overall measure and each of the eight measures and their means which is shown in chart 6.3. This satisfies the **association** validity criteria for empirical validation which states that – a metric has association validity if it has a direct, linear statistical correlation with an external quality factor i.e. the Overall Measure (OM). Though there was no direct relationship of the measures with OM which is needed for **Rank consistency**, figure 36 shows that there is a good amount of correlation between each of the measures and OM (except for LOC).

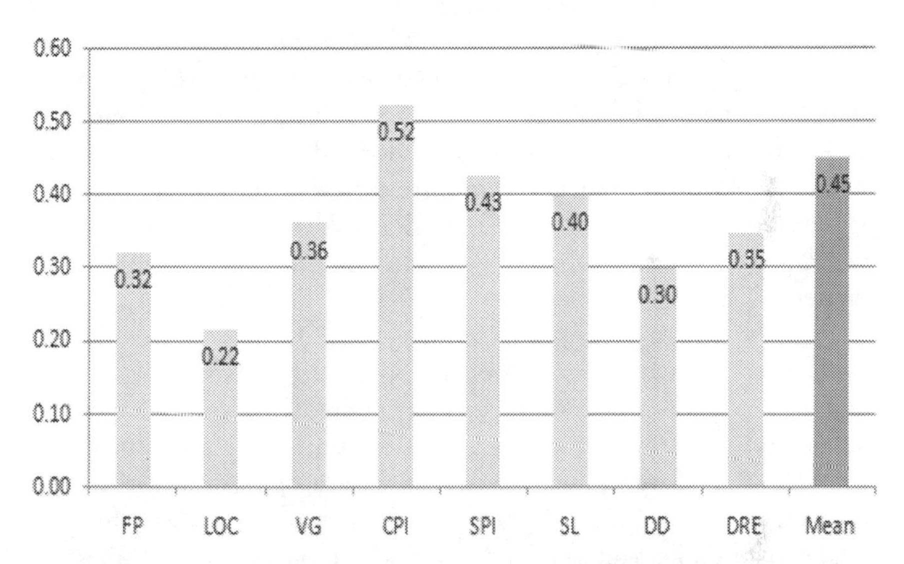

Figure 36: Correlation between the Means of 8 Measures and OM

The Regression model derived was OM (Y) = 2.12 FP + 0.016LOC + 0.24VG + 0.13CPI- 0.107SPI + 0.05SPI + 0.148SL + 0.033DD -0.033DRE with 27.3% of the variability (Value of R-square was 0.273) in the dependent variable (i.e. OM) explained in the regression model. This illustrates a good amount of **Predictability** satisfying another criterion for empirical validation.

The survey data was further analyzed to see if the theoretical constructs brought out a good correlation on some of the related measures. The theoretical constructs that are closely related are the EVM measures – SPI and CPI and the product measures – LOC and v (G). The correlation between SPI and CPI was found to be 0.67 which is expected given that SPI and CPI come from EVM and are often used together as they share a common measurement data infrastructure. However, the correlation between LOC and v (G) was 0.18 which threw a surprise. Further analysis between LOC and v (G) is done in section 6.4 in table 49.

Respondents who rated 1 (strongly disagree) or 2 (disagree) on the measures were categorized as critics. Those who responded with 3 (neither agree nor disagree) were grouped as neutral, while promoters were the ones who rated 4 (agree) or 5 (strongly agree) on the measures. This technique is popularly known as the "Net Promote Score (NPS)" was introduced by Fred Reichheld in 2003. NPS is based on the fundamental perspective that every stakeholder can be divided into three categories: Promoters, Passives/ Neutrals, and Detractors/Critics. The NPS is the difference between the percentage of customers who are Promoters and the percentage who are Detractors.

The most important benefits of the NPS method is simplifying and communicating the objective of creating more "Promoters" and fewer "Detractors" — a concept claimed to be far simpler to understand and act on than more complicated, obscure or hard-to-understand satisfaction metrics or indices. In addition, net promoter method can also reduce the complexity of implementation and analysis, and provide a stable measure of business performance that can be easily compared. An NPS efficiency rating between 50% to 80% is considered good [Reichheld, 2003].

In the survey, 74% of the 110 respondents "promote" that these eight measures can serve as a core set for an objective status of a software project. The complete promoter scores are as shown in figure 37 below.

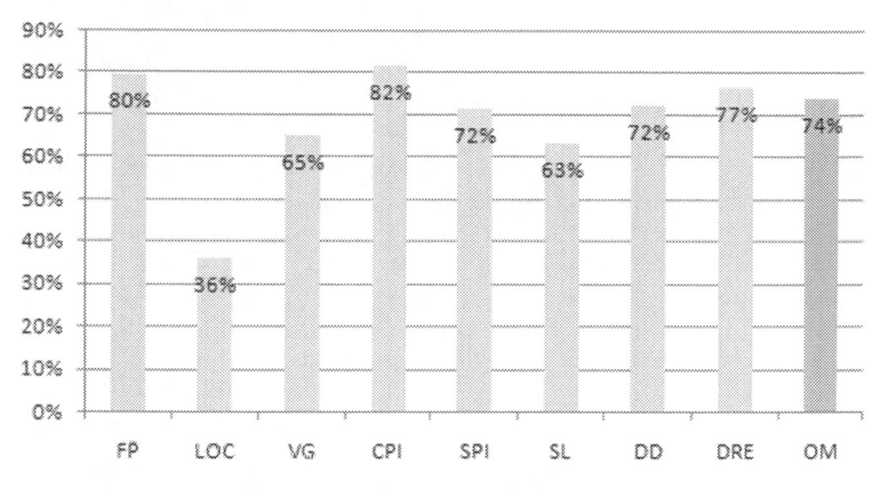

Figure 37: Promoter Responses

The complete critics' scores (without the neutral scores) are as shown in figure 38 below.

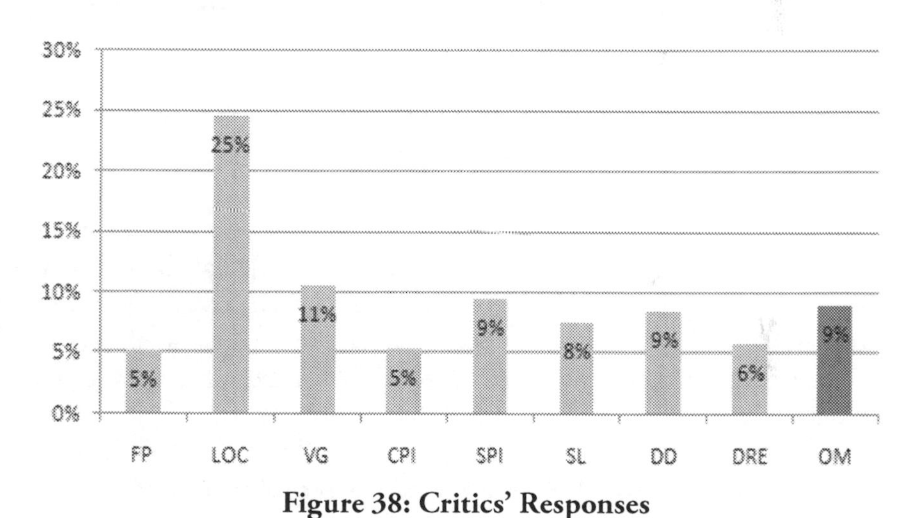

Figure 38: Critics' Responses

To summarize, the survey presents five key pieces of information:

1. The survey addresses four of the six empirical validation criteria. They are repeatability, association, rank consistency and predictability.

2. 74% of the respondents accept that the above eight measures can serve as a core set for an objective status of a software

project. The remaining 26% were either neutral (rated 3 i.e. neutral) or were disagreed (response was 1 or 2) that the eight measures can serve as a core set.

3. Except for LOC, there is a positive correlation between each of the seven measures and the overall measure (OM).
4. The reliability (Index of Variation) is highest for FP at 0.19 and lowest for LOC at 0.30 indicating that the data is reliable.

LOC is the weakest of the eight measures for:

- The Promoter score is at 36%.
- The Pearson's correlation with the OM is at 0.22
- The P-value is 0.001 between LOC and OM data in KW test.

6.4 Debate on having LOC and other measures in the Measurement Framework

While the survey responses were not positive for LOC, the debate on the application of LOC to software measurements has persisted for a long time. There are many academic papers that debate the value of LOC in software engineering. Even in the survey, there were extreme cases where some respondents strongly agreed that LOC measures the size of the software project objectively after development and should be part of the core set of measures, while there were many more who felt otherwise. To gain better understanding, further analysis was exclusively carried out on the LOC relevant survey data.

The first analysis in table 46 was the LOC scores given by respondents of software based organizations and non-software based organization. The analysis shows no discrimination between the types of the organization in relation to this measure.

Table 46: Analysis of LOC data based on Organization types

	Non Software Organizations	**Software Organization**
Average LOC Score	3.00	3.07
Reliability (Index of Variation)	0.34	0.32
Promoters %	32%	37%
Chi-Square Test	P-value is 0.65. As P-value is > 0.05, null hypothesis is accepted implying that the data set is the "same".	
Conclusion	All these measures indicate that there is no difference in the way respondents from non-software organizations and software organizations look at LOC.	

The second analysis in table 47 was the LOC scores given by software project implementers and others (i.e. initiators and beneficiaries). The analysis is as shown in the table 47 indicated little differences in the perception of LOC regardless of the relative role of respondents.

Table 47: Analysis of LOC data based on Stakeholder types

	Implementers	**Others**
Average LOC Score	2.96	3.39
Promoters %	32%	55%
Reliability (Index of Variation)	0.35	0.25
Chi-Square Test	P-value is 0.57. As P-value is > 0.05, null hypothesis is accepted implying that the data set is the "same".	
Conclusion	Though the first three measures show some distinction, the Chi-square test indicates that there is no difference in the way implementers and initiators/ beneficiaries look at LOC.	

The third analysis in table 48 was the LOC scores between developers and non-developers. This was done to see if the developers who write the code can provide any more specific information. The analysis is as shown in the below table shows little distinction between the LOC value as seen by developers and non-developers.

Table 48: Analysis of LOC data based on Developer Skills

	Developers	**Non-Developers**
Average LOC Score	**3.05**	**3.09**
Reliability (Index of Variation)	**0.29**	**0.33**
Promoters %	**28%**	**37%**
Chi-Square Test	P-value is 0.84. As P-value is > 0.05, null hypothesis is accepted implying that the data set is the "same".	
Conclusion	All these measures indicate that there is no difference in the way developers and non-developers look at LOC.	

The fourth analysis in table 49 was between the LOC scores and McCabe's cyclomatic complexity i.e. v (G). As LOC and v (G) are dependent on the programmer's skills, this analysis was done to see if the survey data can throw any interesting information. The analysis is as shown in the table below.

Table 49: Analysis of LOC and v (G)

	LOC	**McCabe's complexity**
Average Score	**3.03**	**3.40**
Reliability (Index of Variation)	**0.34**	**0.30**
Promoters %	**36%**	**65%**

Pearson's Correlation	The Pearson's r for LOC and v (G) data was found to 0.18 which indicates poor correlation between the two measures. As LOC and McCabe's cyclomatic complexity are related, the correlation should have been higher than 0.5.
Chi-Square Test	P-value is 0.80. As P-value is > 0.05, null hypothesis is accepted implying that the data set is the "same" i.e. there is no difference in the way respondents looked at LOC and v (G).
Conclusion	If there was no correlation between the LOC and v (G) data sets, they should have come from different data sets. But as the P-value from the Chi-square test is greater than 0.05, it shows that there is no difference in the data sets .This "anomaly" could be because: LOC is more often used than McCabe's cyclomatic complexity in software projects [Fenton, 2006]. Hence respondents have seen the challenges or failure of using LOC in projects. LOC is simpler to understand than McCabe's cyclomatic complexity for the respondents to give opinion. In addition, due to the technical nature, developers and architects are generally more educated in LOC and v (G) than sponsors and users.

Even with all its problems/deficiencies, LOC is amongst the three measures (along with McCabe's cyclomatic complexity and Function points) that is extensively and routinely used in the industry [Fenton, 2006]. Steve McConnell states "For most organizations, despite its problems, the LOC measure is the workhorse technique for measuring size of past projects and for creating early-in-the-project estimates of new projects. The LOC measure is the *lingua franca* of software estimation, and it is normally a good place to start, as long as you keep its limitations in mind" [McConnell, 2006, pp 199]. Hence LOC is included in the measurement framework with the rider that it should not be the only measure used to make decisions pertaining to project size and should be used along with Function Points.

Some respondents in the survey wanted to have risk and requirements volatility measures in the measurement framework. Risks is a product of two components - Probability of the event happening and Impact/ loss the event can cause. The 'utility' type measure of risk is quite useful for prioritizing risks (the bigger the number the 'greater' the risk) but it is normally meaningless as one cannot get both the probability and the impact numbers. In addition risk exist if there is a challenge in meeting schedule or cost or quality objectives and each of these three attributes are already captured in the measurement framework as SPI, CPI and Cpk/ DD/DRE measures respectively.

Requirements volatility is a measure of how much requirements change after a critical point in the SDLC. It is the ratio of total number of requirements change (add, delete and modify) to total number of requirements for a given period of time. The reason for the change could range from lapse in developer understanding to changes in customer needs as a result of shift in market conditions. This growth/creep/evolution ultimately results in volatility. High levels of requirements volatility ultimately impact the schedule, cost and quality in the project. Each of these three attributes is already captured in the measurement framework as SPI, CPI and Cpk/DD/DRE measures respectively.

It is worth re-emphasizing that, the eight measures proposed are not designed to be the golden measures. Depending on the project needs more measures including requirements volatility and risk can be added or LOC can even be dropped.

6.5 Empirical Validation with a Case study

While a survey is essentially an opinion of the industry practitioners based on their experiences in software projects, we need to see if this framework works in a real world project:

- To continue the second piece of the empirical validation and validate the measurement framework for the remaining two empirical validation criteria i.e. Discriminative power and Trackability.

- To ensure that the implementation challenges in the measurement framework are addressed.

A case study is an in-depth investigation of a single individual, group, or event to find underlying principles within a specific time space. Case studies provide a systematic way of looking at events, collecting data, analyzing information, and reporting the results. This provides a sharpened understanding of why a particular instance happened and what might become important to look more extensively for future research.

In most cases case studies produce much more detailed information than statistical analysis for example survey which was applied in this research. Because its project designs emphasize exploration rather than prescription or prediction, researchers are relatively freer to discover and address issues as they arise in their experiments. Also its **emphasis on context** helps to bridge the gap between abstract research and concrete practice by allowing researchers to compare firsthand observations with the quantitative results obtained through other methods of research.

However, detractors argue that case studies are difficult to generalize because of inherent subjectivity applicable only to a particular context. The approach relies on personal interpretation of data and inferences. Results may not be generalizable, are difficult to test for validity, and rarely offer a problem-solving prescription. Simply put, relying on a few subjects as a basis for cognitive extrapolations runs the risk of inferring too much. [Yin, 2009].

6.6 Case Study Design

There are many types of case studies and selection of a particular case study depends on the goals and/or objectives of the researcher. These different types of case studies are [Yin, 2009]:

1. **Illustrative Case Studies.**

   These are primarily descriptive studies utilizing one or two instances of an event to show what a situation is like. Illustrative case studies serve primarily to make the research topic familiar by proving a common language about the research question.

2. **Exploratory (or pilot) Case Studies.**

   These are condensed case studies performed before implementing a large scale investigation. Their basic function is to help identify questions and select types of measurement prior to the main investigation. The primary pitfall of this type of study is that initial findings may seem convincing enough to be released prematurely as conclusions.

3. **Cumulative Case Studies.**

   These serve aggregate information from several sites collected at different times. The idea behind these studies is the collection of past studies will allow for greater generalization without additional cost or time being expended on new, possibly repetitive studies.

4. **Critical Instance Case Studies.**

   These examine one or more sites for either the purpose of examining a situation of unique interest with little to no interest in generalization. This method is useful for answering cause and effect questions.

The type of case study that is relevant for this research is the "**Critical Instance Case Study" as we have to critically** examine the issues in the implementation/validation of the proposed measurement framework. The pre-requisites however include addressing three key questions [Yin, 2009]:

- **What questions to study?**

  1. The eight core measures (LOC, FP, v (G), SPI, CPI, Cpk, DD and DRE) are the best measures to derive the objective status of their respective attributes.

2. The eight measures can serve as a generic core set for an accurate and objective status of a software project.
3. Can we get complete information to on projects stratus if we use a sub-set of these eight measures?

These three questions can be answered by implementing the measurement framework in controlled and uncontrolled settings.

- **What data are relevant to collect?**

1. Baselined project schedule and budget data including the Earned Value framework.
2. Agreeing on the appropriate CMMI levels for the project. While the organization might be certified in a specific CMMI level, does the project adhere to that standard?
3. The actual data on the eight core measures (LOC, FP, v (G), SPI, CPI, Cpk, DD and DRE).

- **How to analyze that data**

1. The appropriate levels of thresholds, variances and control limits for the eight measures by the stakeholders.

6.7 Case Study 1: Controlled Instance

For a controlled case study, the measurement framework was applied in a SAP Enterprise Portal project for a Canadian bank, whose strategy was to have a Portal for high frequency, customer facing transactions for the tellers, branch managers and back office personnel. The bank has close to 5,000 employees managing assets of over $25 billion and a customer base of about 700,000 people. Since SAP did not have a COTS Portal application for a bank, the project was a bespoke type of project with Java as the programming language accessing the SAP backend banking application with web services. The other key project details were:

- Budget: $2.1 million
- Duration: 1st February to 31st July, 2010.
- Project Implementation team size: 17 people.

- Contracted days: 1,520 person-days (PDs).
- Key project stakeholders: Sponsor, Users, Program Manager, Project Manager, Developer, Testers, Analysts and Business Transformation lead.

The objective of implementing the measurement framework in a controlled setting is to validate the two hypotheses defined in section 3.7.

6.7.1 Metrics Support System

One of the pre-requisites for the measurement program to be successful is to ensure that requirements are complete and non-volatile so that progress can be tracked against a fixed reference a.k.a performance measurement baseline (PMB). So requirements hold the key to the success implementing the measurement framework.

This brings us to the selection of project implementation models. Generally, there are two implementation frameworks applied in software projects: predictive lifecycle and adaptive software development. Each one differs in their approach to the software development lifecycle (SDLC). The predictive lifecycle favors optimization over adaptability, whereas adaptive software development is much more flexible while welcoming change. Incrementalism and iteration which are part of the adaptive (or agile) development help a project include new values and new challenges discovered along the way. They also help flush out implicit requirements while we can still address them. According to Capers Jones adaptive software development works well for 1,000 FP range software projects [Jones, 2002].

At the same time a fundamental principle in software projects is changes introduced late in the project lifecycle costs far more than changes were introduced earlier in the cycle [Davis, Alan M, Bersoff, Edward and Comer, Edward, 1988]. So use of the Waterfall model (i.e. predictive lifecycle) seeks to reduce the overall cost and risk by moving as much of the thinking up front as possible. Hence to get requirements completely (and be measured against the fixed target i.e. PMB) at an optimal cost and time, the best practices from both predictive lifecycle and adaptive software development methodologies were combined into a "hybrid methodology" in this project as shown in the figure 39 below.

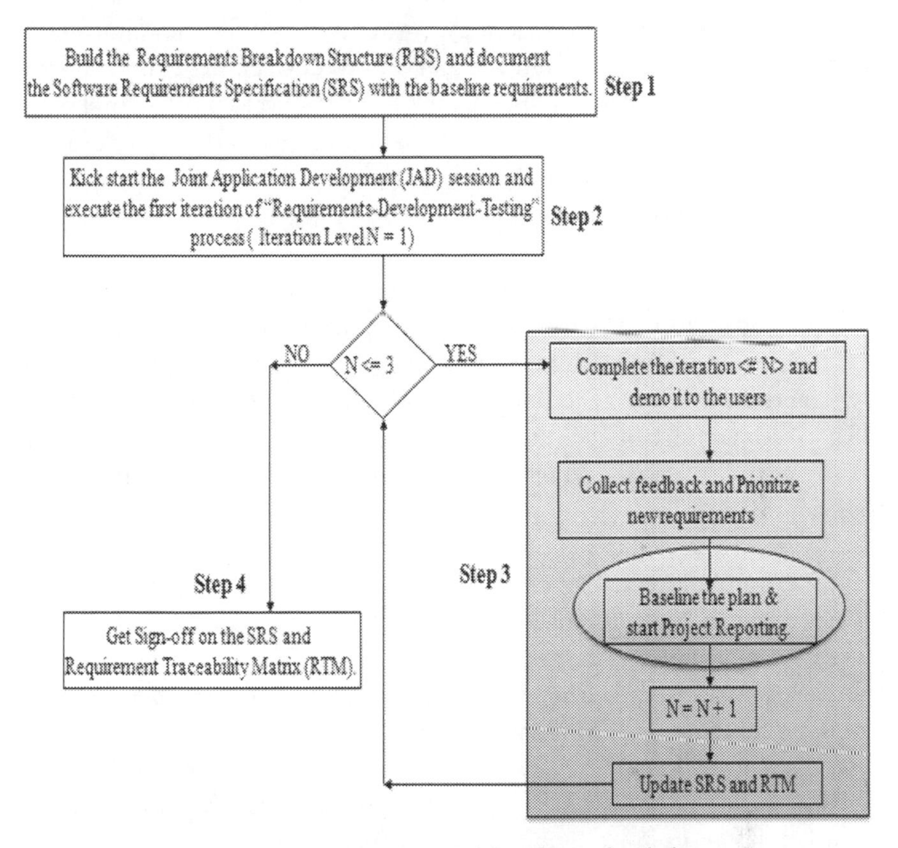

Figure 39: Four Step Hybrid Methodology

When the first iteration of the "Requirements-Development-Testing" activity was completed (box circled in figure 39), the project status reporting using the eight measures was started.

6.7.1.1 Size and Complexity Measures

The first measure to capture was the **LOC**. This was a straight forward task as the Java standard framework used for the Portal development provided this value directly.

To determine the **v (G),** a freeware called "Eclipse plug-in" was used. This was a freeware downloaded from the internet; http://eclipse-metrics.sourceforge.net. This tool calculates the McCabe's cyclomatic complexity along with various measures which provided more details on the health of the code base. The output from executing this tool for the "Transfers" function (one of the modules in the SAP Portal development) is as shown in the figure 40 below where the total McCabe's cyclomatic complexity for the module is the sum of the maximum McCabe's cyclomatic complexity for the individual packages such as TransfersComp.java, RecurringTransfersView.java etc.

From the v (G) metric, the every method was categorized as low, medium and high complexity as per the tables provided by Frappier et al [Frappier et al, 1994] and endorsed by SEI. The complexity from every method was aggregated at the module level. From this complexity categorization and LOC count, the FP Languages table provided by QSM Inc [QSM, 2009] was leveraged to determine the number of **FPs** in the application.

For example, after the first iteration the v (G) for Transfers functionality is 1,380 and the LOC is 18,603. As the cyclomatic complexity is more than 50, this functionality was categorized as highly complex. For a highly complex functionality with Java as the programming language, the FP languages table from QSM recommends 214 Lines of Code per FP [QSM, 2009].Hence the number of FPs in the Transfers component = 18,603/214 = 87. So using the Eclipse plug-in toll two metrics were determined i.e. v (G) (directly) and FP (indirectly).

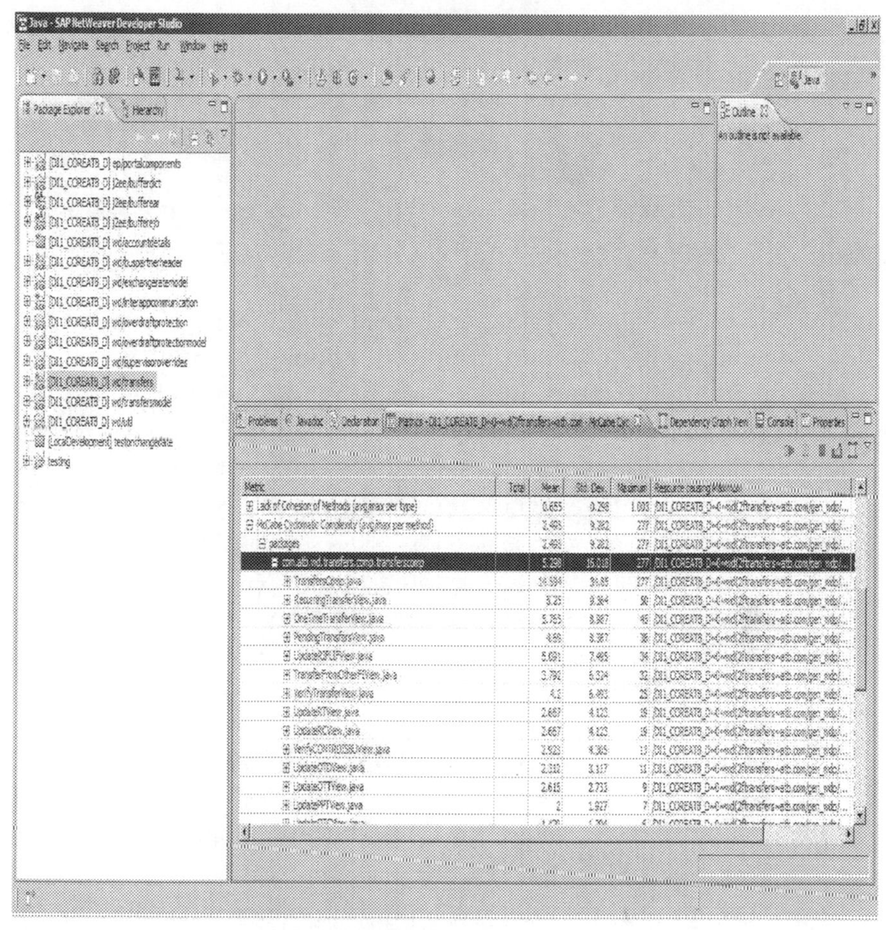

Figure 40: McCabe's Output from Eclipse plug-in

6.7.1.2 Schedule and Cost Measures

Both the **SPI and CPI** were derived from EVM principles using Microsoft Project 2003 [Microsoft, 2011]. The three important steps used for calculating the SPI and CPI were:

1. **Building the Work Breakdown Structure (WBS)**

   Using the SRS document, a list of mutually exclusive elements were identified as the derivation of SPI and CPI for Earned Value (EV) analysis was to be done where the task i.e. work package (lowest element in the WBS) would be assigned to the developer. Each of the four main activities i.e. requirements, design, development and testing including deployment in the SDLC was a work package and the aggregation of the four respective work packages was the control account (CA).

2. **Assigning Planned Values (PV) to each terminal WBS element i.e. Control Account**

   After building the WBS, the planned value (PV) was assigned to each terminal element i.e. work package of the WBS by preparing software estimates such as high level effort (in person-days), project duration and resource cost. PV is the total cost of the work planned and is the product of Rate (cost of accomplishing the task by a resource) and Total Planned duration. With the base lined SRS (Step 1), the development was started with high level effort and the PV. Later when the project was re-baselined (first iteration of the "Requirements-Development-Testing"), the efforts were suitably changed in Microsoft Project for more accuracy.

3. **Establishing Earning Rules for each terminal WBS element**

   The values of SPI and CPI are dependent on the "earning rules" which quantify the work accomplished i.e. Earned Value (EV) for each terminal element of the WBS. The EV compares the actual work accomplished to the planned work up to a specific point in time. For this project the earning rules were defined as shown below in table 50 considering incremental and iterative development.

Table 50: Earning Rules for SPI and CPI

Deliverable	Percentage Applied for Earning Rules				
	25	50	75	90	100
RTM #	Technical design (TDD) is completed and development started	1st iteration of development completed and feeedback elicited from the stakeholders	2nd iteration of development completed and feeedback elicited from the stakeholders	3rd iteration of Development completed, testing is also completed and all defects are fixed	All 3 iterations are completed. Buffer consumed. SRS and TDD completed and signed-off

6.7.1.3 Quality Measures

The results pertaining to the three measures on quality are explained below.

1. To calculate the **Cpk** every test case was considered as an opportunity for a defect to potentially surface. For example 105 out of 117 test cases executed failed in the Transfers functionality at the first iteration. This translated to 897,436 Defects for 1 million opportunities (DPMO) providing a Sigma Level of 0.2.

2. **DD** was used to compare the number of defects against the number of FPs in each software component. In the Transfers functionality, 105 defects were identified for 87 FPs giving a DD of 105/87 = 1.21

3. **DRE** is the ratio of the defects removed at a specific phase to the total possible defects. It provides information on the rate at which defects are resolved. As explained in chapter 3, to find the potential count of the total defects in a new software project, Capers Jones proposes to use FPs to the power 1.25 [Jones, 2002]. For example, the total number of defects that

can potentially exist in the Transfers functionality was FPs to the power 1.25 i.e. 87 to the power 1.25 which is 265. After the first iteration was completed, 12 defects were fixed giving a DRE of 12/265 = 0.05.

The complete measurement framework implemented in the case study is as shown in figure.41 below.

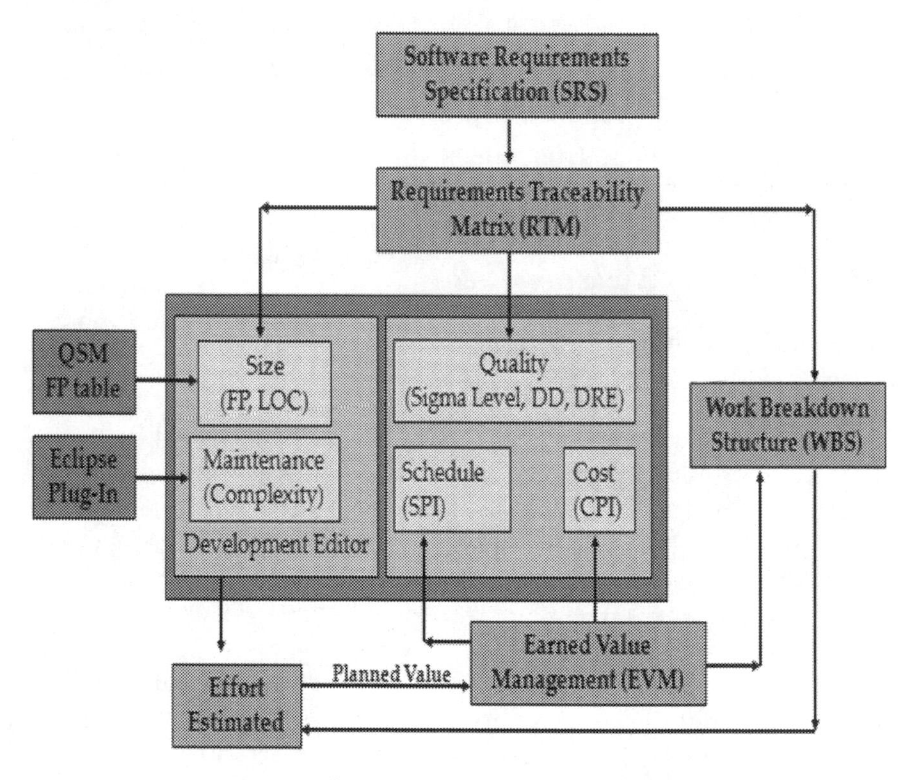

Figure 41: Measurement Framework

6.7.2 Analysis of Case Study Results

The first project status report using the measurement framework was presented when the first iteration (the circled box in figure 39) was completed. The status report is as shown in the table 51 below.

Table 51: Measurement Data after the First Iteration

Project Status as on May 15th, 2010 (End of First Iteration)																
Sl #	FDD Name	Lines of Code (LOC)	Function Points (FP)	Total McCabe Complexity	Complexity Type	CPI	SPI	Total Test Cases (Opportunities)	Defects Open incl untested Test Cases	DPMO	Sigma Level	DD	Total Defects possible	Defects Removed	DRE
1	PT_EH_01-Deposit & FDD Maintenance	2154	10	295	High	1.50	0.81	34	31	911765	0.15	3.08	18	3	0.17
2	PT_EH_01-LO-Deposit & FDD Maintenance - Change A/c Holder	1067	5	160	High	0.73	1.00	27	13	481481	1.55	2.61	27	14	0.52
3	PT_EH_04-Registered Plans & TFSA - Contract Details	2764	13	341	High	0.83	0.81	10	6	600000	1.25	0.46	24	4	0.16
4	PT_EH_05-Registered Plans & TFSA - Contract Maintenance	1728	8	371	High	1.00	0.94	24	20	833333	0.53	2.48	24	4	0.17
5	PT_EH_09-Loan - Account Details	865	4	117	High	1.00	0.92	20	4	200000	2.61	0.99	20	16	0.80
6	PT_EH_10_16_18-Loan - Make Payment + Lumpsum + Payout	2521	12	346	High	0.80	1.05	15	13	866667	0.39	1.10	22	2	0.09
7	PT_EH_14-Loan - Disbursement - Loan Maintenance	2064	10	367	High	1.00	1.03	20	16	800000	0.66	1.66	20	4	0.20
8	PT_EH_17-Loan - Payout Scenerio - Loan Maintenance	1513	7	273	High	1.18	1.12	10	2	200000	2.34	0.28	12	8	0.69
9	PT_EH_20-View Estatements	1307	6	238	High	1.40	1.00	35	30	857143	4.02	4.91	35	5	0.14
10	PT_EH_24-Transfers	18603	87	1380	High	0.76	0.86	117	105	897436	0.23	1.21	265	12	0.05
11	PT_EH_26-Fees	2298	11	229	High	0.91	0.91	45	37	822222	0.58	3.45	45	8	0.18
12	PT_EH_27-Supervisor Overrides	1032	5	90	High	1.42	0.87	8	7	875000	0.35	1.45	7	1	0.14
13	PT_EH_32_33-Overdraft Protection Details & Maintenance	4471	21	641	High	1.53	0.85	45	41	911111	0.15	1.96	45	4	0.09
14	PT_EH_60-Deposit & FDD Account Details	1298	6	172	High	1.49	0.96	7	1	142857	2.57	0.16	10	6	0.63
15	PT_EH_61-Registered Plan Payments incl Close Accounts	10107	47	1216	High	1.59	0.97	100	91	910000	0.16	1.93	124	9	0.07
16	PT_EH_62-Deposits - Close Accounts	1620	8	197	High	1.69	1.00	26	14	538462	2.57	1.85	26	12	0.46
17	PT_EH_64-FDD Funding for Non Registred Products	2094	10	271	High	0.69	0.87	17	14	823529	0.57	1.43	17	3	0.17
	Overall	37506	269	6704	High	1.01	0.94	560	445	794643	0.68	1.66	741	115	0.16

The final project status report is as shown in the table 52 below.

Table 52: Measurement Data after the Final Iteration

Project Status as on July 31st, 2010 (Project Ending; All three iterations completed)																
Sl #	FDD Name	Lines of Code (LOC)	Function Points (FP)	Total McCabe Complexity	Complexity Type	CPI	SPI	Total Test Cases (Opportunities)	Defects Open incl unresolved Test Cases	DPMO	Sigma Level	DD	Total Defects	Defects Removed	DRE
1	PT_EH_01-Deposit & FDD Maintenance	2390	11	309	High	1.15	1.00	34	2	58824	3.07	0.18	34	32	0.94
2	PT_EH_01-LO-Deposit & FDD Maintenance - Change A/c Holder	1195	6	155	High	0.79	1.00	27	2	74074	2.95	0.36	27	25	0.93
3	PT_EH_04-Registered Plans & TFSA - Contract Details	2988	14	341	High	0.68	1.00	10	1	100000	2.78	0.07	27	7	0.26
4	PT_EH_05-Registered Plans & TFSA - Contract Maintenance	1793	8	394	High	1.11	1.00	24	2	83333	2.88	0.24	24	19	0.79
5	PT_EH_09-Loan - Account Details	971	5	126	High	1.30	1.00	20	1	50000	3.15	0.22	20	18	0.90
6	PT_EH_10_16_18-Loan - Make Payment + Lumpsum + Payout	2600	12	370	High	0.86	1.00	15	2	133333	2.61	0.16	23	14	0.62
7	PT_EH_14-Loan - Disbursement - Loan Maintenance	2183	10	353	High	1.00	1.00	20	8	400000	1.75	0.78	20	12	0.60
8	PT_EH_17-Loan - Payout Scenerio - Loan Maintenance	1543	7	291	High	0.96	1.00	10	3	300000	2.02	0.42	12	7	0.59
9	PT_EH_20-View Estatements	1307	6	243	High	1.63	1.00	35	3	85714	2.87	0.49	35	32	0.91
10	PT_EH_24-Transfers	18841	88	1494	High	1.10	1.00	117	2	17094	3.62	0.02	270	115	0.43
11	PT_EH_26-Fees	2335	11	241	High	0.95	1.00	45	12	266667	2.12	1.10	45	33	0.73
12	PT_EH_27-Supervisor Overrides	1167	5	104	High	1.35	1.00	32	12	375000	1.82	2.20	32	20	0.63
13	PT_EH_32_33-Overdraft Protection Details & Maintenance	4982	23	687	High	1.30	1.00	45	9	200000	2.34	0.39	51	33	0.65
14	PT_EH_60-Deposit & FDD Account Details	1419	7	183	High	1.65	1.00	7	1	142857	2.57	0.15	11	6	0.56
15	PT_EH_61-Registered Plan Payments incl Close Accounts	10724	50	1426	High	1.30	1.00	120	25	208333	2.31	0.50	133	95	0.71
16	PT_EH_62-Deposits - Close Accounts	1643	8	212	High	1.87	1.00	26	2	76923	2.93	0.26	26	24	0.92
17	PT_EH_64-FDD Funding for Non Registred Products	2241	10	289	High	1.00	1.00	17	1	58824	3.07	0.10	19	15	0.80
	Overall	60322	282	7218	High	1.04	1.00	624	88	141026	2.58	0.31	808	490	0.61

The eight measures were used to regularly present the project status every week. This weekly status reporting helped the stakeholders track the project objectively and make corrective actions on scheduling (Fast Tracking, Crashing and Resource Leveling), budgeting, buffer management etc for successful project completion. Measuring the reliability of the measurement data/instrument was unnecessary because the framework was objective relying on one data infrastructure. Some of the corrective actions taken in the project based on the eight measures were:

- Increased testing and assigning highly skilled developers (with focus on effective defect resolution) for bigger sized and more complexity functions. For example, Transfers functionality based on the first iteration report.
- Functions with poor DRE and low SPI (Example is PT_EH_04 – Registered Plans & TFSA – Contract Details where the DRE was 0.09 and SPI was 1.05 after the first iteration) functions were put on the "Max-Attention" list.
- Functions with low sigma levels and low complexity/size/SPI were prioritized due to their "Low-hanging fruit" status.
- Impact analysis was primarily driven from FPs. For instance after every iteration of "Requirements-Development-Testing" for every increase/decrease of one FP, approximately 125 LOC were also factored and other measures were derived for complete impact.

Figure 42 below shows improvement in the overall project status between the first status report and the final status report.

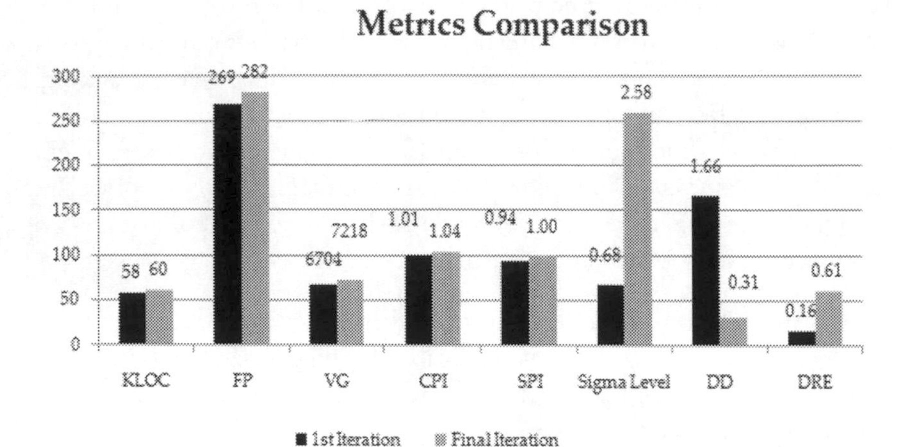

Figure 42: Comparison of the Eight Measures

Significant improvements are on Cpk, DD and DRE metrics indicating improvements in the project quality. The increase in Cpk shows improvement in the SDLC process due to reduced defects. The reduction in DD indicates increase in product stability while the increase in DRE shows improvement in the rate at which defects are resolved. This satisfies the two remaining criteria for empirical validation:

- **Discriminative power.** For the eight measures, the critical threshold value for SPI and CPI is anything less than one, while the project benchmarks for Cpk was at 2.0, DD was at 0.5 and DRE was at 0.5.

- **Trackability.** This criterion is used to assess whether a component is improving, degrading, or stagnating in quality over time. For example, Cpk, DD and DRE are shown to be trackable, as fixing the defects increases the quality in the project.

As mentioned in the hypothesis, the primary success criteria for the **initiators and implementers** is that - a successful project is one that is completed within 10% of its committed cost and schedule and has delivered all of its intended functions [Humphrey, 2005]. In this project, the tracking was against the base-lined plan for which the metrics were presented after the first iteration of "Requirements-Development-Testing" was completed.

Accordingly, the projected schedule at completion = Initial Duration/SPI = 6/0.94 = 6.38 Months and the projected cost at completion = Initial Budgeted Cost/CPI = 1520/1.01 = 1504 Person-days (PDs). When the project was completed on 31st July, the schedule at Completion = Initial Duration/SPI = 6/1.00 = 6 Months while the Cost at Completion = Initial Budgeted Cost/ CPI = 1520 /1.04 = 1462 PDs. So both the schedule and cost were within 10% of the committed cost and schedule for the agreed scope.

In addition the hypothesis states that, the primary success criteria for the **beneficiaries** are based on two parameters.

a. Defect rate according to CMMI levels.

This project was managed according to CMMI level 3 standards (i.e. defined business process) where the standards, process descriptions, and procedures were tailored from the organization's set of standard processes to suit this particular project. For a CMMI level 3 organization the defect rate should be 0.27 defects per FP [Kan, 2003]. This translates to 77 defects for the 282 FPs in scope for this project (77 = 282 * 0.27) and this number i.e. 77 is close to the 88 defects still open to be resolved as shown in table 52. So as far as the user's primary concern of quality is concerned, the existing defect count when the project is completed meets their requirements as it is close to SEI CMMI Level 3 standards.

b. DRE levels

According to David Longstreet, a software project is mature if the DRE is greater than 45% [Longstreet, 2008] and this project meets the standard as the DRE was 61%.

It is not practical to bring the defect count to zero in any project and this always invites discussion if it is feasible to achieve such a goal. In this project, these two defect criteria values (the defect rate of 88 defects and DRE of 61%) were acceptable to all stakeholders in the project as the strategy was to deploy it along with other dependent application on a particular date. Moreover the must-have functionalities were defect free and regression tested.

Apart from this, the measurement framework also served as a basis for clear and objective communication with project stakeholders, promoted teamwork, improved team morale as individual team member efforts were tied to the project goals.

6.8 Case Study 2: Uncontrolled Instance

To demonstrate a cause and effect hypothesis, the case study should also indicate that a phenomenon occurs after a certain treatment is given to a subject (controlled setting), and that the phenomenon does not occur in the absence of the treatment (uncontrolled setting).

While the eight measures that were applied in the SAP Portal project were in a controlled setting, another similar project (Internet Banking application built on Java with Web services) was studied in the same program at the same time in an uncontrolled setting. The eight measures were not applied and instead the project tracking was done with earned value management using just the SPI and CPI measures which is one of the most popular techniques for tracking project performance today.

On 15th May 2010, the SPI for this project was at 0.96 with the planned completion date of 30th July 2010. On 9th July, the SPI was at 0.99 still showing the same planned completion date of 30th July. But on 30th July, when the project was expected to be completed (and transitioned to the maintenance team), there were many defects that were still open making the project incomplete. After recalibration, the project completion date was then moved to 20th November ultimately resulting in an unsuccessful project as per the operationalization of the hypothesis described in chapter 2.

Two prominent issues come up with the usage of the SPI as a performance metrics in this case study.

1. Traditionally, EVM tracks schedule variances not in units of time, but in units of currency (e.g. dollars) or sometimes quantity (e.g. person days) considering that it is more natural to speak of schedule performance in units of time. SPI or more specifically schedule variance (SV) says project is $500 000 late. But it doesn't provide information on how many months or days is the delay. In addition, an unfavorable

schedule variance does not necessarily imply that project is behind schedule as by itself, the SPI reveals no critical path information. The SPI should be used in conjunction with other schedule information [Fleming, 1992].

2. As mentioned before, SPI is the ratio between EV and PV. At the end of the project, if the project has delivered everything that was planned, EV and PV must be equal making SPI equal to 1. This characteristic without accounting for the deadline, questions the usefulness and predictability of SPI because after the planned end of the project, PV remains constant while the EV will continue to grow until the actual end of the project. For example, if the project is planned to be finished in 10 months, and if the EV after 10 months is 70%, it is 30 % behind schedule. Finally at the end of the project (as eventually the project has to be completed) the EV will be equal to your PV when all the deliverables are completed. This according to EVM means that the project is completed "on time" even though it is couple of months late. Even worse is if the project "continues" after the planned end date and for some reason if nothing is done say for three months, the SPI will be the same for three months as it still refers to the baselined PV/duration. In addition, a project can languish near completion (e.g. SPI = 0.95) and never be flagged as outside acceptable numerical tolerance. This is exactly what has happened in this case study where SPI was used to track the project performance.

Certain researchers maintain that SPI is good and reliable in first two thirds of the project and it starts to be defective over the final third of a project's lifecycle [Lipke, 2003; Vandevoorde, S. and Vanhoucke M, 2006].To correct these problems, Earned schedule (ES) proposed by Lipke [Lipke, 2003] looks promising. The idea of ES is analogous to EV; instead of cost, time would be used for measuring project performance. ES is an emerging practice though it is not the standardized and recommended methodology to measure project schedule. This is outside the scope of current research and would be explored in future research as an extension of this research thesis.

In addition, while EVM measures indicated that the project was on schedule, it did not portray the true project status as the quality (or product) measures particularly on defect prevention and removal were never reported and hence no corrective action was taken. If the weekly status reports had the quality measures in them (which was deteriorating every week), perhaps the project stakeholders would have been better prepared to take suitable corrective actions.

Moreover in the survey, respondents preferred CPI (promoter score of 82%) over SPI (promoter score of 72%) again indicating that though CPI and SPI come from EVM, SPI is not as popular as CPI.

6.9 Lessons Learned from Case Studies

While the objective of the case studies (controlled and uncontrolled) was primarily empirically validation, the case studies threw some interesting light on the implementation aspects of measurement frameworks. More than a technical exercise, the implementation of the measurement framework was an organizational initiative involving people, process and technology. Below are some important lessons learned during the implementation of this measurement framework.

- As measures are meant for tracking the project delivery by the team, train and educate the project team on how the project progress is tracked for their inclusiveness in the measurement program.
- When making decisions with these measures, consider the trend over multiple time periods.
- Be sensitive to the fact that improving one measure may cause another measure to deteriorate.
- Use one standard format of the status report and assign the responsibility of the status report to the Project Manager as multiple formats from different sources lead to multiple interpretations.

- Recognize the business situation surrounding the measures and use measures to get the direction needed to accomplish the project goals.
- Understand the "culture" and "acceptance" of these measures in the project.
- Though metrics are expensive and time consuming to define, collect, analyze and report; only the first status report takes significant time. The subsequent status reports take much less time.
- Finally, be sensitive to people's aspirations and their situations in the project. Do not measure or compare people with measures. Metrics simply provide the foundation and rationale for decision making.

6.10 Other Empirical Validation Criteria

As discussed in chapter 2, empirical validation included eight criteria. In additional to the six empirical validation criteria of Schneidewind discussed in the survey and two case studies, there are two more empirical validation criteria namely Instrument Validity and Attribute Validity.

1. **Instrument Validity**

The instrument validity tells us how to capture the data we seek. A metric has instrument validity if the underlying measurement instrument is valid and properly calibrated. For example, consider a tool that has not been calibrated. As a result, the data gathered from that tool would be considered invalid. In this backdrop, instrument validity is already addressed in Kaner-bond questionnaire i.e. question #5.

2. **Attribute Validity**

Attribute validity is whether the attribute is actually exhibited by the measured entity. Kitchenham et al's discussion of attribute validity focuses on the actual measurements of a measure. They make the point that the attributes being measured need to be aspects of software that have both intuitive and well-understood meanings. For example, if one were assessing the validity of attribute "psychological complexity" of code, one might poll a group of developers and empirically evaluate their agreement. The attribute validity for the eight measures is as shown in table 53.

Table 53: Attribute Validity

Measure	**Attribute**	**Attribute Validity**
LOC	Physical program size; size after development	**Yes**
FP	Functional program size; size before development	**Yes**
v(G)	Program Complexity	**Yes**
CPI	Cost/Budget Performance; Productivity	**Yes**
SPI	Schedule Performance	**Yes**
Cpk	Process Stability	**Yes**
DD	Module Stability	**Yes**
DRE	Quality Achieving Velocity	**Yes**

Both the measure and the attribute have to satisfy dimensional analysis. Dimensional analysis checks relations among physical quantities by using their dimensions and is based on the fact that a physical law must be independent of the units used to measure the physical variables.

6.11 Conclusion

Apart from satisfying the eight empirical validation criteria, the two types of empirical validations (Survey and case study) have provided additional information as shown below in table 54.

Table 54: Summary of Empirical Validation

Survey Results									
Parameter	FP	LOC	VG	CPI	SPI	Cpk	DD	DRE	OM
Acceptability by Practitioners	High	Low	High	High	High	High	High	High	High
Relation to the Overall Measure (OM)	High	Low	High	High	High	High	High	High	NA
Case Study Results									
Parameter	FP	LOC	VG	CPI	SPI	Cpk	DD	DRE	OM
Degree of Objectivity	Medium	High	High	Low	Low	Medium	Medium	Medium	Medium
Ease of Derivation	Medium	Low	Low	Medium	Medium	Medium	Medium	Medium	Medium
Level of Accuracy	Medium	High	High	Medium	Medium	Medium	Medium	Medium	High
Complexity in Implementation	Medium	Low	Low	Medium	Medium	Medium	Medium	Medium	Medium

This chapter shows the validation of the generic GQM framework using the feedback from seasoned software professionals and real world project experience. This is reflected in the empirical validation of the measures that have been assessed, statistical analysis conducted and patterns found. . The survey and the two case studies strengthen the hypothesis empirically that the proposed measurement framework when properly implemented provided critical information at the right time for proactive decision making. However, this research is still not free of any limitations. Chapter 7 looks at the conditions under which the measurement framework will best work including directions for future research.

Chapter 7: Conclusion and Directions for Future Research

7.1 Overview

Software project management involves making critical decisions based on the severity of different parameters at different SDLC phases as shown in the figure 43 below.

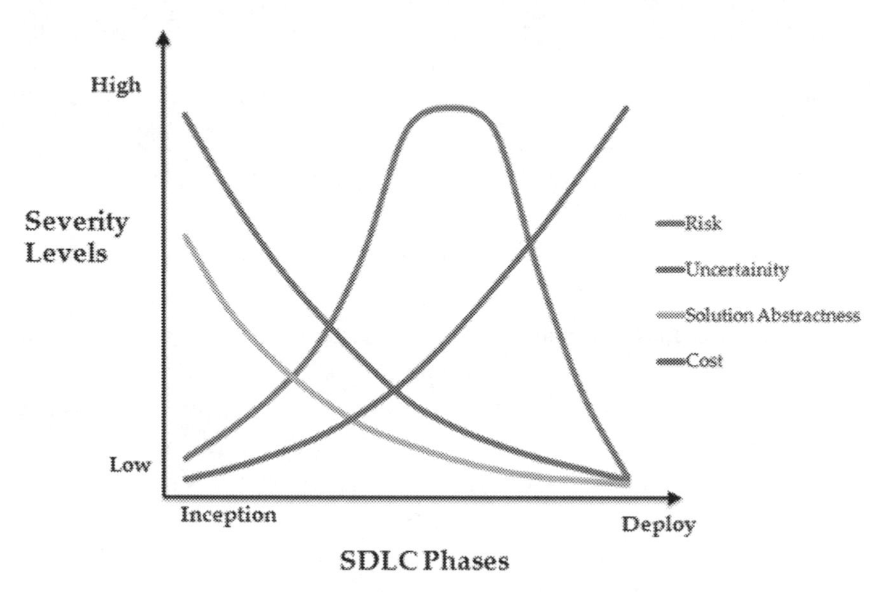

Figure 43: Software Project Criticality

Measurement frameworks which identify key events and trends are invaluable in guiding software organizations in making informed decisions. Applying measurement frameworks in software project is an amalgamation of art and science. The "art" dimension encompasses human intuition, communication, and capabilities to negotiate conflicting objectives and constraints. The "science" factor refers to formalizing the problem and applying measurement theory axioms to generate best possible measures for effective empirical validation and acceptance by practitioners.

7.2 Conditions for Application

This measurement framework is software process improvement (SPI) initiative following the four tenets of software process improvement (SPI) [Kuntzmann, 1997].

i. Evolution is possible and it requires time and resources.
ii. At higher process maturity levels, risk decreases and performance increases.
iii. Evolution implies the existence of a pre-defined sequence to keep processes under control.
iv. Organizational maturity will decrease if it is not maintained over time.

Hence in addition to the CSFs outlined in section 3.9, the two case studies have provided the below list of additional factors which will increase the possibility of successfully implementing the proposed measurement framework.

7.2.1 Product Factors

- **Project Type**. As mentioned in the hypothesis section in chapter 2, the measures for the software project are for Bespoke and Hybrid-COTS development projects. These measures are not designed for software maintenance or COTS type of projects.

- **Size.** As FPs calculation is labor-intensive and expensive for applications greater than 15,000 FPs [Jones, 2002], this measurement framework works well for applications less than 15,000 FPs. (An application greater than 15,000 FPs can be categorized as a program and broken down into smaller projects for better control and management.)

7.2.2 Process Factors

- **Realistic PMB**. Requirements should be complete, correct and non-volatile. Based on this, the effort estimate (for planned value in EVM) should be done rationally and the project should be baselined for the measurements can be carried out against this baseline.
- **Mature data collection infrastructure**. Given that measurement programs are expensive and time consuming to define, collect, report and analyze the measurement data collection infrastructure (Tools and Methodology) should be available for the data to be collected regularly and made readily available at all times. This is generally found in SEI CMMI level 3 organizations and higher.
- **Regular Application**. Hence the measurement framework should be applied regularly and frequently in every phase of the SDLC. This improves the feedback process and ultimately the project delivery.

7.2.3 Resource Factors

- **Stakeholder commitment**. The project stakeholders should be committed to the measurement program. Every stakeholder from the developer to the tester to the senior management should actively participate in the measurement process to make it successful.

7.3 Future Research

Future research on the measurement framework can involve work in the following areas for increased reliability of the measurement framework:

1. GQM+
2. Requirements engineering
3. Effort estimation
4. EVM Earning rules
5. Cognitive Sciences
6. Other types of projects

7.3.1 Research Area 1- GQM+

The success of a measurement initiative in a software company depends on the quality of the links between metrics programs and organizational business goals. However in chapter 3, we started with the derivation of the measures with the project goal s of the stakeholders assuming that the project goals of stakeholders are aligned to the organizational business goals. However this is an assumption and one of the disadvantages of GQM paradigm could be the misalignment of organizational business goals and project measures. In this circumstance, GQM+ strategies is an extension to GQM which adds the capability to create measurement programs that ensure alignment between business goals and strategies, software specific goals and measurement.

7.3.2 Research Area 2- Requirements Engineering

A dominant reason for failure in a software project has been attributed to requirements as requirements drive almost every project activity in the SDLC such as scoping, estimating/ budgeting, scheduling, coding, testing, documentation and training. "Clear Statement of Requirements" is at #3 in the list of critical factors from Standish Group that make software projects successful [Standish, 2009]. A controlled study by SEI found that the most important factor for project success was the way the project team did its "requirements elicitation and management" [Woody, 2005]. Hence a measurement framework to specifically address the challenges in the requirements phase will help the stakeholders to take remedial actions early in the project.

7.3.3 Research Area 3 - Effort Estimation

The Chaos summary 2009 report shows a substantial increase in both cost and time overruns. Cost overruns increased from 47% in 2006 to 54% in 2008 and Time overruns have gone up from 72% in 2006 to 79% in 2008 [Standish, 2009]. Though there are costs incurred in a project such as software installation, training and travel costs etc..., bulk of the cost in a software project is accounted towards human effort which is less deterministic compared to other types of costs. In addition to this the EVM measures – SPI and CPI are dependent on the planned value (PV) and the PV in turn is dependent on the effort estimated. But accurately estimating the PV is a significant challenge in software projects because:

1. Uncertainty in the size of the project requirements
2. Estimation model to apply. Though many quantitative software cost estimation models have been developed such as COCOMO, SLIM (Software Lifecycle Model) there is no clear and a definite answer on which software estimation model to use.
3. Bias towards underestimation.
4. Management pressure for lower estimates to win the deal/ project.
5. Estimating the productivity of the project team as effort is a function of size and productivity.

7.3.4 Research Area 4 - Earning Rules

The standard earning rules in EVM are closely associated with the NDIA (National Defense Industrial Association) ANSI/EIA (*American National Standards Institute* /Electronics Industries Alliance)-748 standards which contain 32 criteria drawn from five process areas [Budd and Budd, 2005]. These earning rules and criteria are more suited for a production floor set up where the relationship between output and input is linear, predictable and visible unlike software projects. The earning rules and the criteria need to be revalidated for sufficiency, appropriateness and relevancy for software projects.

7.3.5 Research Area 5 – Cognitive Sciences in Software Projects

Complexity in a project exists in product, process and resources. Some are stand alone while many more interact with each other. While the product complexity is addressed with McCabe's cyclomatic complexity, to understand the process and resource complexity the research has to encompass cognitive sciences (an interdisciplinary scientific study of information concerning faculties such as perception, language, reasoning, and emotion).

7.3.6 Research Area 6 - Other types of Software Projects

This measurement framework is not applicable for COTS based systems such as SAP ERP, Oracle ebusiness suite, Seibel CRM etc. The science of measuring COTS based systems (CBS) has not yet advanced to the point where there are standard measurement methods and only a few enterprises regularly measure COTS quality [Cechich and Piattini, 2004].

Another potential area of improvisation of this measurement framework is towards software maintenance projects. Software maintenance involves making changes to the software after delivery in order to correct faults and deficiencies found during usage, or to improve performance or any other attribute as well as to adapt the product to a new environment.

7.4 Conclusion

Science thrives on measurement and in any scientific field measurement generates quantitative descriptions of key products, processes and resources. With appropriate measurement framework we can clarify and understand the issues better and take suitable corrective actions.

This enhanced understanding helps to select better techniques and tools to control and improve the software project delivery. Lord Kelvin said,"If you cannot measure something, then you do not understand it". Hence a good measurement framework not only provides the information for improvement but also acts like a catalyst in improving the project visibility, and eventually attaining the project goals.

Good software project management demands only a handful of measures which are regularly applied to gauge the health of a project. The eight measures are meant to be minimum-set to increase the odds of knowing the project status holistically and to complete the project on time and budget. Apart from providing vital information on the project performance, the measurement framework can also serve as the basis for clear and objective communication with project stakeholders, promote teamwork and improve team morale by linking efforts of individual team members to the project goals. Finally, there is no engineering without measurement and software engineering can become a true engineering discipline if we build a solid foundation of measurement-based theories [Pfleeger et al, 1997].

References

- Aggarwal KK and Singh Yogesh, 2007, "Software Engineering", 3rd edition, New Age International Publishers, New Delhi.
- Al-Ahmad, Walid, Al-Fagih, Khalid, Khanfar, Khalid, Alsamara, Khalid, Abuleil, Saleem, Abu-Salem, Hani, 2009, "A Taxonomy of an IT Project Failure: Root Causes", International Management Review, Volume 5, Number 1.
- Albrecht, Alan, October 1979, "Measuring Application Development Productivity", Proceedings of the Joint SHARE, GUIDE, and IBM Application Development Symposium, Monterey, CA.
- Allen, Mary J and Yen, Wendy M, 2001, "Introduction to Measurement Theory ", Waveland Pr Inc.
- Alreck, P. L., & Settle, R. B, 1995, "The survey research handbook: Guidelines and strategies for conducting a survey", 2nd Edition, Burr Ridge, IL: Irwin.
- Ambler, Scott , May 22, 2008, "Dr. Dobb's Agile Newsletter 05/08", http://www.drdobbs.com/architecture-and-design/207801786;jsessionid=C0NYT4TCILUAHQE1GHOSKH4ATMY32JVN
- Atkinson, Roger, 1999, "Project management: cost, time and quality, two best guesses and a phenomenon, it's time to accept other success criteria", International Journal of Project Management Vol. 17, No. 6, pp. 337 to 342.
- Bache, R M and Neil M D, 1995, "Introducing Metrics into Industry: A Perspective on GQM. In Software Quality Assurance and Measurement: A Worldwide Perspective. Edited by N. Fenton, R. Whitty and Y. Iizuka. International Thomson Computer Press, London.
- Basili, V, 2003, "Matching Software Measurements to Business Goals", http://www.cs.umd.edu/~basili/presentations/2003%20SMM-ASM%20Keynote.pdf

- Basili, V and Boehm, B, May 2001, "COTS-Based Systems Top 10 List", IEEE Xplore, Volume 34, Issue 5, 91-95.
- Basili, Victor R, Gianluigi Caldiera, and Rombach, H. Dieter, 1994, "The Goal Question Metric Approach", Encyclopedia of Software Engineering, Wiley.
- Basli, Victor, Lindvall, Mikael, Regardie, Myrna, Seaman, Carolyn, Heidrich, Jens, Munch, Jurgen, Rombach, Deiter, Trendowicz, Adam, 2007, "Bridging the gap between business strategy and software development", 28th International Conference on Information Systems, Montreal, Canada.
- Batty, Michael and Torrens, Paul, 2005, "Modeling and prediction in a complex world", Futures, 37(7), pp: 745-766.
- Berander, Patrik and Per, Jönsson, September 21–22, 2006. "A Goal Question Metric Based Approach for Efficient Measurement Framework Definition", ISESE'06, Rio de Janeiro, Brazil.
- Boehm, B.W, 1981, "Software Engineering Economics", Englewood Cliffs: Prentice Hall
- Breiman L., Friedman J.H., Olshen R.A., Stone C.J., 1984, "Classification and Regression Trees", Belmont CA. Wadsworth International Group.
- Briand, Lionel, Morasca, Sandro and Basili Victor, 1994, "Property based Software Engineering Measurement", Internal Report 94.078, Politecnico di Milano, Dipartimento di Elettronicae Informazione.
- Briand, Lionel, El Emam, Khaled and Morasca, Sandro, 1995, "Theoretical and Empirical Validation of Software Product Measures", International Software Engineering Research Network.
- Briand, Lionel, El Emam, Khaled and Morasca, Sandro, 1996, "On the Application of Measurement Theory in Software Engineering", International Software Engineering Research Network (ISERN).
- Brooks, Frederick, 1995, "The Mythical Man-Month", 20th Edition, Pearson Education.
- Budd, Charles and Budd, Charlene, 2005, "A practical guide to Earned Value Project Management", Management Concepts.

- Card, D.N and Jones, C.L, November 2003, "Status report: Practical Software Measurement", Proceedings of the third International Conference on Quality Software, pp 315-320.
- Cechich, Alejandra and Piattini, Mario, 2004, "On the measurement of COTS functional suitability", Lecture Notes in Computer Science, Volume 2959/2004.
- Chen, Kung H and Shimerda, Thomas A , Spring 1981, "An Empirical Analysis of Useful Financial Ratios", Financial Management, Vol. 10, No. 1, pp. 51-60
- Chidamber, S.R. and Kemerer, C.P., June 1994, "A Metrics Suite for Object Oriented Design", IEEE Trans Software Eng., vol. 20, no. 6, pp. 476-493.
- Christian, David S and Ferns, Daniel V, 1995, "Using earned value for performance measurement on Software Development projects." Acquisition Review Quarterly.
- Clemmer, Jim, 1992, "Firing on All Cylinders", pp. 264, TCG Press.
- Cooper, Donald R and Schindler Pamela S, August 2000, "Business Research Methods", McGraw-Hill College; 7th edition.
- Crosby, Philip B, 1995, "Quality Without Tears: The Art of Hassle-Free Management", pp. 59, McGraw-Hill Professional.
- Damascvicius, Robertas and Stuikys, Vytautas, December 2010,"Metrics for Evaluation of Metaprogram Complexity", Computer Science and Information Systems (ComSIS) Journal Vol. 7, 770 No. 4.
- Dalcher Darren, October 2009, "Software Project Success: Moving Beyond Failure", The European Journal of Informatics Professional, Volume 10, Issue 5.
- Davies , Terry Cooke, 2002, "The "real" success factors on projects", International Journal of Project Management, Volume 20, pp 185–190
- Davis, Alan M, Bersoff, Edward and Comer, Edward, October 1988,"A Strategy for Comparing Alternative Software Development Life Cycle Models", IEEE transaction on Software Engineering, Volume 14, Number 10.
- Dawson, Catharine, 2002, "Practical Research Methods", UBS Publishers, New Delhi.

- Dekkers, Carol and McQuaid, Patricia, Mar/Apr 2000, "The Dangers of Using Software Metrics to (Mis) Manage", IT Professional, vol. 4 no. 2, pp. 24-30.
- DeMarco, Tom, 1986, "Controlling Software Projects: Management, Measurement, and Estimates", Facsimile edition, Prentice Hall.
- DeMarco, Tom, 1995, "Why Does Software Cost So Much?" Dorset House Publishing Company.
- Drucker, Peter F, 2001, "The Essential Drucker", 1st edition, Harper Business.
- Dwayne, Phillips, 1998, "The Software Project Managers Handbook", IEEE Computer Society.
- El Emam, K, June 2000, "A Methodology for Validating Software Product Metrics", National Research Council of Canada, NCR-ERC-1076.
- Far, Behrouz Homayoun, 2008, "Software Metrics", http://enel.ucalgary.ca/People/far/Lectures/SENG421/PDF/SENG421-01.pdf
- Fenton, E Norman and Pfleeger, SL, 1997, "Software Metrics: A Rigorous and Practical Approach", PWS Publishing Company.
- Fenton, E Norman and Neil, Martin, 1999, "Software metrics: successes, failures, and new directions", Journal of Systems and Software.
- Fenton, E Norman, September/October 1999, "A Critique of Software Defect Prediction Models", IEEE transactions on Software Engineering, Volume 25, Number 6.
- Fenton, E Norman, September 2006, "New Directions for Software Metrics", CIO Symposium on Software Best Practices.
- Fitsilis, Panos, March 2009, "Measuring the complexity of software projects", Computer Science and Information Engineering, 2009 WRI World Congress on, vol 7, pp 644-648.
- Fleming, Q. W, 1992, "Cost/Schedule Control Systems Criteria", Probus Publishing, Chicago, IL.
- Fleming, Quentin W and Koppelman, Joel M, July 1998. "Earned Value Management – A Powerful Tool for Software Projects", Journal of Defense Software Engineering.

- Fleming, Quentin W and Koppelman, Joel M, 2000,"Earned Value Project Management, Second Edition", PM World Today, Volume 8, Issue 8.
- Fleming, Quentin W and Koppelman, Joel M, 2006,"Start with simple Earned Value on all your projects", Project Management Institute.
- Florac William A, Park, Robert E and Carleton Anita D, April 1997, "Practical Software Measurement: Measuring for Process Management and Improvement", Software Engineering Institute.
- Fiammante Marc , April 2010 , "Managing the complexity of business processes" http://www-01.ibm.com/software/solutions/soa/newsletter/apr10/article_complex_bus_processes.html
- Forrester (Narsu, Uttam), November 2003, "Defect Removal Efficiency Should Be Your Key Quality Metric for Applications", "http://www.forrester.com/rb/Research/defect_removal_efficiency_should_be_key_quality_metric_for_applications/q/id/33143/t/2"
- Frappier, M., Matwin, S., Mili, A, 1994, "Software Metrics for Predicting Maintainability. Software Metrics Study", Tech. Memo. 2. Canadian Space Agency.
- Gaffney, JR, 1984, "Estimating the Number of Faults in Code," IEEE Transactions of Software Eng., vol. 10, no. 4.
- Gardy, Robert , 1992, "Practical software metrics for project management and process improvement", Prentice Hall
- Gartner, October 2008, "Gartner Identifies Four Disruptions That Will Transform the Software Industry", http://www.gartner.com/it/page.jsp?id=778112
- Gido, Jack and Clements, James P, 1998,"Successful Project Management", South-Western Pub.
- Goodman, Paul, 2004, "Software metrics: best practices for successful IT management", pp 6, Rothstein Associates Inc.
- Gray, Martha M, November-December 1999, "Applicability of Metrology to Information Technology", Journal of Research of the National Institute of Standards and Technology, Volume 104, Number 6.

- Gross, Joshua B, April 2006, “End User Software Engineering: Auditing the Invisible”, WEUSE II Workshop, Montréal, Quebec, Canada.
- Hall, Tracy and Fenton, Norman, Mar/Apr. 1997, “Implementing Effective Software Metrics Programs,” IEEE Software, vol. 14, no. 2, pp. 55-65.
- Hanna, Robert A, 2009,”Earned Value Management Software Projects”, Third IEEE International Conference on Space Mission Challenges for Information Technology.
- Harry, Mikel, 2000, “Six Sigma: The Breakthrough Management Strategy Revolutionizing the World’s Top Corporations”, Doubleday.
- Hatton, Les, May 1996, “Programming Research Ltd”, IEEE Software.
- Hetzel, C.W, 1993, “Making software measurement work”, QED.
- Hofmann, Hubert F and Lehner, Franz, July/Aug 2001, “Requirements Engineering as a Success Factor in Software Projects”, IEEE Software, vol. 18, no. 4, pp. 58-66.
- Hubbard, Douglas (2010), “How to Measure Anything: Finding the Value of “Intangibles” in Business“, 2nd Edition, Wiley.
- Humphrey Watts S, March 2005, “Why Big Software Projects Fail: The 12 key Questions”, Journal of Defense Software Engineering.
- Humphrey, Watts S, 1995, “A Discipline for Software Engineering”, Reading, MA: Addison-Wesley.
- Humphrey, Watts, 1999, “Managing the Software Process”, Pearson Education.
- IEEE, 1998, “1061-1998 - IEEE Standard for a Software Quality Metrics Methodology”.
- IEEE, 1990, “610.12-1990 - IEEE Standard Glossary of Software Engineering Terminology”
- IFPUG, 1999, Function Point Counting Practices Manual Release 4.1,
- Jones, Capers, 1995, Keynote Address, 5th International Conference of Software Quality, Austin, TX.
- Jones, Capers, 1996, Applied Software Measurement: Assuring Productivity and Quality, McGraw-Hill.

- Jones, Capers, 2002, "Software Estimating Rules of Thumb", IEEE Software, vol. 29, no. 3, pp. 116.
- Kan, Stephen, 2003, "Metrics and Models in Software Quality Engineering", 2nd Edition, Pearson Education.
- Kerzner, Harold, 2003, "Project Management: A Systems Approach to Planning, Scheduling, and Controlling", 8th Edition, Wiley.
- Kaner, Cem and Bond, Walter P, 2004, "Software Engineering Metrics: What Do They Measure and How Do We Know", 10th International Software Metrics Symposium.
- Kaplan, R S and Norton, D P, Jan – Feb 1992, "The balanced scorecard: measures that drive performance", Harvard Business Review, pp. 71–80.
- Kitchenham, Barbara, 1993, "Making Sense of Metrics", Research debate by Software Reliability and Metrics Club at Glasgow.
- Kitchenham, Barbara, Pfleeger, SL and Fenton, Norman, December 1995, "Towards a Framework for Software Measurement Validation", IEEE Transactions on Software Engineering, Volume 21 Issue 12.
- Kothari C R, 2001, "Research Methodology: Methods and Techniques", New Age International.
- Kriz,Jürgen , 1988, "Facts and Artifacts in Social Science: An Ephistemological and Methodological Analysis of Empirical Social Science", McGraw Hill Research, New York, NY, USA.
- Kuntzmann, Combelles A, August 1997, "The Business performance perspective", ESSI g-r-a-m, No 3, pp 4.
- Kurekova, Eva, 2001, "Measurement Process Capability – Trends and Approaches", Measurement Science Review, Volume 1.
- Lehman, M. M, J. F. Ramil, P. D. Wernick, D. E. Perry and W. M. Turski , 1997, "Metrics and laws of software evolution—the nineties view", Proc. 4th International Software Metrics Symposium (METRICS '97), pp. 20-32.
- Linberg, Kurt R, "Software developer perceptions about software project failure: A case study", Journal of Systems and Software. Vol. 49, no. 2, pp. 177-192. 1999.

- Lipke, WH, 2003 summer, "Schedule is Different", Measurable News.
- Longstreet David, 2008, "Test Cases and Defects", http://www.softwaremetrics.com/Articles/defects.htm
- Lukas, Joseph, 2008, "Earned Value Analysis – Why it doesn't work", AACE International Transactions
- Mauricio J. Ordonez, Hisham M. Haddad, 2008, "The State of Metrics in Software Industry," Fifth International Conference on Information Technology: New Generations, pp. 453-458.
- McCabe, Thomas, December 1976, "A Complexity Measure", IEEE Transactions on Software Engineering: 308–320.
- McConnell, Steve, 1993, "From Anarchy to Optimizing", "http://www.stevemcconnell.com/articles/art02.htm".
- McConnell, Steve, 2006, "Software Estimation: Demystifying the Black Art", Microsoft Press.
- McConnell, Steve, May/June 1997, "Gauging Software Readiness with Defect Tracking", IEEE Software, Vol. 14, No. 3.
- McGaghie, William C, Bordage, Georges, Crandall, Sonia and Pangaro, Louis, September 2001, "Reseracxh Design", Academic Medicine, Volume 76, Issue 9, pp 929-930
- McManus, John and Harper Trevor Wood, "Management issues accounted for 65% of causal factors identified with failed projects", Management Services, Autumn 2007, pp 39-43.
- Meneely Andrew, Smith Ben, Williams, Laurie, 2010, "Software Metrics Validation Criteria – A Systematic Literature review", Transactions on Software Engineering Methodologies.
- Microsoft Project 2003, 2011, http://office.microsoft.com/en-au/project-help/applying-earned-value-analysis-to-your-project-HA001021179.aspx.
- Miller, G.A, 1957, "The magical number 7 plus or minus two: Some limits on our capacity for processing information", Psychological Review, Volume 63, pp. 81-97.
- Mills, Everald, 1988, "Software Metrics", SEI Curriculum Module SEI-CM-12-1.1, Carnegie Melon Software Engineering Institute.

- Moller KH and Paulish DJ, 1993, "Software metrics: A practitioner's guide to improved product development", 1st Edition, IEEE Computer Society Press
- Muketha, G.M, Ghani AAA, Selamat MH and Atan R, 2010, "A Survey of Business Process Complexity Metrics", Information Technology Journal, Volume 9, pp 1336-1344.
- Ordonez Mauricio J and Haddad Hisham M, 2008, "The State of Metrics in Software Industry," Fifth International Conference on Information Technology: New Generations, pp. 453-458.
- Park Robert E, Goethert Wolfhart B and Florac William A, August 1996, "Goal-Driven Software Measurement—A Guidebook", Software Engineering Institute.
- Park, Robert E. et al, September 1992, "Software Size Measurement: A Framework for Counting Source Statements", Software Engineering Institute, Carnegie Mellon University.
- Parviz F. Rad, Ginger Levin, 2006, "Metrics for project management: formalized approaches", Management Concepts.
- Paul Ssemaluulu and Ddembe Williams, 2007, "Complexity and Risk in IS Projects: System Dynamics Approach".
- Pfleeger SL, Ross Jeffery, Bill Curtis, Barbara Kitchenham, 2002, "Status Report on Software Measurement", IEEE Software, Volume 14, Issue 2, pp 33-43.
- Pfleeger, SL, Nov /Dec. 2008, "Software Metrics: Progress after 25 Years?", IEEE Software, vol. 25, no. 6, pp. 32-34.
- Pfleeger, SL Jeffery, Ross, Curtis, Bill and Kitchenham, Barbara, Mar/Apr 1997, " Status report on software measurement" IEEE Software, vol. 14, Issue 2, pp. 332-43.
- Philliber, Susan Gustavus, Schwab, Mary Bast, and Sloss, Sam, 1980, Peacock Publishers, Inc.
- PMI, 2004, "Practice Standard for Earned Value Management", Project Management Institute.
- PMI, 2008, "PMBOK 4th Edition", Project Management Institute.
- Polgar, Stephen and Thomas, Shane A, 2000, "Introduction to Research in Health Sciences", Churchill Livingstone; 4 edition.

- Pressman, Roger, 2004, "Software Engineering: A Practitioner's Approach", 6th Edition, McGraw-Hill publications.
- Putnam, Larry and Meyers, Ware, August 2002, "Control the Software Beast With Metrics-Based Management", The Journal of Defense Software Engineering.
- Pyzdek, Thomas, 2000."The Six Sigma Handbook - 1st Edition", McGraw-Hill Companies.
- QSM, November 2009, "Function Point Languages Table-Version 4.0". http://www.qsm.com/?q=resources/function-point-languages-table/index.html
- Rad, Parviz F and Levin, Ginger, September 2005, "Metrics for Project Management: Formalized Approaches", Management Concepts; 1 edition.
- Reichheld, Frederick F, Dec 2003, "One Number You Need to Grow", Harvard Business Review.
- Reel, John S, May/June 1999, "Critical Success Factors in Software Projects", IEEE Software, Volume 16 Issue 3, May 1999
- Remington Kaye and Pollack, Julien Feb 2008, "Tools for Complex Projects", Gower Publishing
- Remus, H, 1983, "Integrated Software Validation in the View of Inspections/Review", Proceedings of the Symposium on Software Validation, pp 57-64.
- Sauer, Chris, Gemino, Andrew and Horner Reich, Blaize, November 2007, "The impact of size and volatility on IT project performance", Communications of the ACM, Volume 50 Issue 11.
- Schneidewind, Norman F, May 1992, "Methodology for Validating Software Metrics", IEEE Transactions on Software Engineering, Volume 18, Number 5.
- Schwalbe Kathy, "IT Project Management", 4th Edition, Thomson Course Technology, 2004.
- Sessions, Roger, March 2010, "IT Complexity Crisis; Danger and Opportunity", Fourth International Workshop on Software Quality and Maintainability. Madrid, Spain.
- SEI, 1997, C4 Software Technology Reference Guide, pp 147.

- Shepperd, M. Kadoda, G, November 2001, “Comparing software prediction techniques using simulation”, IEEE Transactions on Software Engineering, Volume 27 Issue 11.
- SIIA (Software & Information Industry Association), 2001, “Software as a Service: Strategic Backgrounder”.
- Simon, Phil, 2009, “Why New Systems Fail: Theory and Practice Collide”, AuthorHouse; 1st edition
- Sirvio, Komi, 2003, “Development and Evaluation of Software Process Improvement Methods – Doctoral Dissertation”, VTT Publications.
- Software Tech News, 2010, “Goal-Question-Metric (GQM) approach” https://goldpractice.thedacs.org/practices/gqm/index.php, Volume 13, # 3.
- Soni Devpriya, Shrivastava Ritu and Kumar M, 2009, “A Framework for Validation of Object Oriented Design Metrics”, International Journal of Computer Science and Information Security, Vol. 6, No.3.
- Standish Group, 2009, “Chaos Summary 2009 Report”.
- Symons, Charles, 2001, “Come Back Function Point Analysis (Modernized) – All is Forgiven!)”, Proc. of the 4th European Conference on Software Measurement and ICT Control, FESMADASMA, Germany, 2001, pp. 413-426.
- Tomayko, James and Hallman, Harvey, July 1989, “Software Project Management - SEI Curriculum Module SEI-CM-21-1.0”, Software Engineering Institute.
- Tatikonda , M.V, Rosenthal S.R, Feb 2000, “Technology novelty, project complexity, and product development project execution success: a deeper look at task uncertainty in product innovation”, IEEE Transactions on Engineering Management, Volume: 47 Issue:1, pp 74 - 87.
- Tausworthe, Robert C, 1980, “The Work Breakdown structure in Software Project Management”, Journal of Systems and Software.
- Tjaden, Gary S, 1996, “Measuring The Information Age Business”, Technology Analysis and Strategic Management, Volume 8, Number 3, pp. 233-246.
- Trochim, William M.K., 2002, “Construct Validity”, http://trochim.human.cornell.edu/kb/constval.htm

- Turner, J.R, Zolin, R and Remington, K, October, 2009, "Modeling Success on Complex Projects:Multiple Perspectives over Multiple Time Frames", Proceedings of IRNOP IX, Berlin.
- Vandevoorde, S and Vanhoucke, M, 2006, "A comparison of different Project Duration Forecasting Methods using Earned Value Metrics", International Journal of Project Management.
- Vierimaa, M, Tihinen, M, Kurvinen, T, May 2001, "Comprehensive approach to software measurement", Proc. of the 4th European Conference on Software Measurement and ICT Control, FESMA-DASMA 2001, Heidelberg, Germany, pp 237-246.
- Van Solingen, Rini and Berghout, Egon, "The Goal/Question/Metric Method: A Practical Guide for Quality Improvement of Software Development", McGraw Hill, 1999
- Wang, YingXu, 2003, "The Measurement Theory for Software Engineering", IEEE CCECE, pp 1321 - 1324 vol.2
- Ward, Stephen and Chapman, Chris, February 2003, "Transforming project risk management into project uncertainty management", International Journal of Project Management, Volume 21, Issue 2, Pages 97-105.
- Weidong Xia and Gwanhoo Lee, 2005, "Complexity of Information Systems Development Projects: Conceptualization and Measurement Development", Journal of Management Information Systems, Vol. 22, No. 1, pp. 45-83.
- Westfall, Linda, 2003, "Are We Doing Well, Or Are We Doing Poorly?", http://www.westfallteam.com/Papers/Are_We_Doing_Well.pdf
- Westfall, Linda, 2005, "12 steps to Useful Software Metrics", http://www.westfallteam.com/Papers/12_steps_paper.pdf
- Wiegers,Karl,2002,"21 Project Management Success Tips", http://www.ieee.or.com/Archive/21project_management_tips.pdf"
- Wiegers, Karl, 2003, "Software Requirements Engineering", Software Requirements, 2nd Edition , Microsoft Press
- Wilkens, Tammo T, 1999, "Earned Value, Clear and Simple", http://www.acq.osd.mil/pm/old/paperpres/wilkins_art.pdf

- Wohlin Claes et al, 2000, Experimentation in Software Engineering, 2000, Kluwer Academic Publishers Norwell.
- Wolverton, R.W., June 1974, "The Cost of Developing Large-Scale Software:, IEEE Transactions on Computers, Volume C-23, No 6, pp 615-636.
- Woody, Carol, March 2005, "Eliciting and Analyzing Quality Requirements: Management Influences on Software Quality Requirements", Technical Note CMU/SEI-2005-TN-010
- Yin, Robert K, 2009, "Case Study Research: Design and Methods", Fourth Edition, SAGE Publications, California.
- Zuse, Horst, 1997, "Foundations of Software Measurement", Walter de Gruyter Publications.

Index

A

B

C

D

E

F

G

H

T

U

V

W

Appendix 1: Glossary of Terms

- **Baseline Plan**. A baseline plan is a point of reference. The plan used as the comparison point for project control reporting. There are three baselines in a project—schedule baseline, cost baseline and scope baseline. The combination of these is referred to as the performance measurement baseline.

- **Empirical software engineering.** It is a sub-domain of software engineering focusing on experiments on software systems (software products, processes, and resources). It is interested in devising experiments on software, in collecting data from these experiments, and in devising laws and theories from this data.

- **Earned value management (EVM)**. It is a project management technique for measuring project performance and progress in an objective manner combining measurements of scope, schedule, and cost in a single integrated system.

- **Goal-Question-Metric (GQM)**. GQM is a top-down approach to establish a goal-driven measurement system for software development, in that the team starts with organizational goals, defines measurement goals, poses questions to address the goals, and identifies metrics that provide answers to the questions.

- **Measurement theory**. Measurement theory is a branch of applied mathematics that is useful in measurement and data analysis. The fundamental idea of measurement theory is that measurements are not the same as the attribute being measured. Hence, if you want to draw conclusions about the attribute, the nature of the correspondence between the attribute and the measurements has to be considered.

- **Reliability.** It refers to the consistency of a measure. A test is considered reliable if we get the same result repeatedly.
- **Software Development Lifecycle (SDLC).** The systems development lifecycle (SDLC) is a conceptual model used in managing software projects that describes the stages involved in an information system development project, from an initial feasibility study through maintenance of the completed application.
- **Validity.** Validity is the extent to which a test measures what it claims to measure. It is vital for a test to be valid in order for the results to be accurately applied and interpreted. Validity is not determined by a single statistic, but by a body of research that demonstrates the relationship between the test and the behavior it is intended to measure.

Appendix 2: Survey Questionnaire

The participants were requested to rate the eight measures in the below set of nine questions on a scale of 0 to 5 as follows.

0 - Never heard of this Measure

1 – Strongly Disagree

2 – Disagree

3 – Neither Agree nor Disagree

4 – Agree

5 – Strongly Agree

The nine questions are:

1. Can **Function Points (FPs)** measure the size of the software project objectively before development?
2. Can **Lines of Code (LOC)** measure the size of the software project objectively after development?
3. Can **McCabe's Cyclomatic Complexity (VG)** measure the complexity of the software project objectively for software maintenance?
4. Can **Cost Performance Index (CPI)** measure/track the cost/effort of the software project objectively?
5. Can **Schedule Performance Index (SPI) determine**/track the schedule of the software project objectively?

6. Can **Sigma level (Cpk)** measure the quality of the software development process in the project objectively?
7. Can **Defect Density (DD)** measure the quality of one software component against another in the project?
8. Can **Defect Removal Efficiency (DRE)** measure the efficiency of defects resolved in the project?
9. Can these **8 Core measures in totality (OM)** fairly describe the accurate and objective status of the software project?

About the Author

Prashanth Harish Southekal draws close to 15 years of software engineering experience working for companies such as Accenture, SAP AG and General Electric in North America, Europe and India. He has consistently delivered successful small, medium, large-sized cross-functional software projects from $5k to $10 million in size for the Engineering, Automotive and Banking sectors. He has spoken at various software conferences including the IEEE, IAMOT, ERP-Expert and SAP- ASUG conferences. He holds PhD in Software Engineering, MS in Information Technology and BS in Mechanical Engineering. He was a regular faculty of Lean Six Sigma (LSS) certification program in General Electric and has conducted numerous half day workshops titled "Juggling for Creativity" and "Six Thinking Hats" for leading software organizations in Bangalore, India. He lives in Calgary, Canada with his wife Shruthi and two children – Pranathi and Prathik. He can be reached at prashanth_sh@yahoo.com

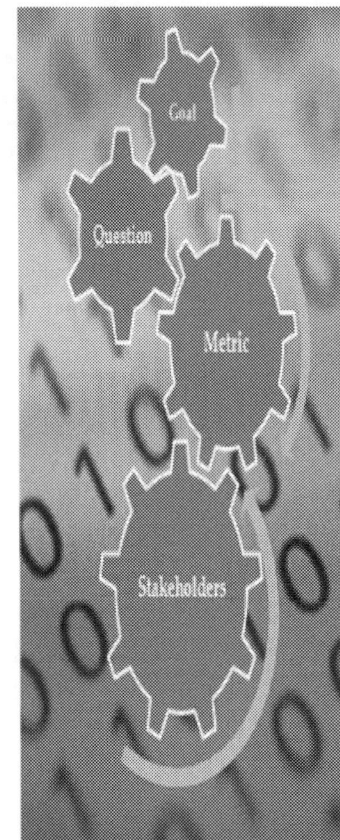

ABOUT THE BOOK

Software projects today are often characterized by poor quality, schedule overruns and high costs. Hence project decision makers need an objective and validated measurement framework to allocate limited resources and to track project progress. In this backdrop, based on the Goal-Question-Metric (GQM) model, Prashanth Harish Southekal has come up with eight generic objective measures for the project stakeholders to base their corrective actions for successful project delivery . The measurement framework is validated (i) theoretically with measurement theory criteria and (ii) empirically with case studies (Controlled and Uncontrolled) including a global survey representing industry practitioners from 29 countries.

ABOUT THE AUTHOR

Prashanth Harish Southekal draws close to 15 years of software engineering experience working for companies such as Accenture, SAP AG and General Electric. He has spoken at various software conferences including the IEEE, IAMOT, ERP-Expert and SAP-ASUG conferences. He holds PhD in Software Engineering, MS in Information Technology and BS in Mechanical Engineering.